Advances in Boundary Lubrication

Advances in Boundary Lubrication

Guest Editor

William Chong

MDPI

Basel • Beijing • Wuhan • Barcelona • Belgrade • Novi Sad • Cluj • Manchester

Guest Editor
William Chong
Institute for Sustainable
Transport
Universiti Teknologi Malaysia
Johor
Malaysia

Editorial Office
MDPI AG
Grosspeteranlage 5
4052 Basel, Switzerland

This is a reprint of the Special Issue, published open access by the journal *Lubricants* (ISSN 2075-4442), freely accessible at: https://www.mdpi.com/journal/lubricants/special_issues/35Z5124044.

For citation purposes, cite each article independently as indicated on the article page online and as indicated below:

Lastname, A.A.; Lastname, B.B. Article Title. *Journal Name* **Year**, *Volume Number*, Page Range.

ISBN 978-3-7258-5205-5 (Hbk)
ISBN 978-3-7258-5206-2 (PDF)
https://doi.org/10.3390/books978-3-7258-5206-2

Contents

About the Editor

William Chong

William Chong is affiliated with the Institute for Sustainable Transport (IST-HICoE) at Universiti Teknologi Malaysia. His research concentrates on tribology, lubrication science, and sustainable transport systems, with a recent focus on hydrogen internal combustion engines, the synthesis and application of bio-based lubricants, and the incorporation of circular economy principles into mobility solutions. His work further encompasses waste valorisation and thermal conversion processes, particularly in developing sustainable feedstocks for advanced lubricants and alternative fuels. He has led and participated in numerous international collaborative projects, including extensive research programmes on diesel–hydrogen dual-fuel engines, lubricant durability in advanced powertrains, and sustainable lubricant formulation. His accomplishments include authoring articles in leading scientific journals, serving on advisory and editorial boards of international publications, and actively engaging in academia–industry collaborations to support next-generation engine and lubricant technologies. Through his endeavours, he aims to bridge fundamental tribological science with practical engineering applications, thereby promoting sustainability and technological innovation within the transport sector.

Preface

This Reprint presents a collection of research articles focused on advancing boundary lubrication, an essential area for reducing friction, wear, and energy loss in mechanical systems. It includes molecular-level simulations, predictive models, experimental techniques, and the development of advanced lubricants and additives, such as nanomaterials and sustainable bio-based formulations. The aim is to offer a comprehensive overview of how fundamental understanding and applied research together improve tribological performance under challenging operating conditions.

The motivation for compiling this Reprint stems from the urgent need for reliable and sustainable lubrication solutions that promote efficiency, durability, and environmental responsibility in engineering applications. By collecting contributions from various fields within tribology, this collection acts not only as a record of current advances but also as a guide for future research directions.

This work is intended for researchers, engineers, and students in tribology, mechanical engineering, and materials science, as well as industry practitioners, for whom advanced lubrication strategies are essential. We hope that this compilation will inspire new ideas, foster interdisciplinary collaboration, and support the ongoing advancement of boundary lubrication science and technology.

William Chong
Guest Editor

lubricants MDPI

Article

Coarse-Grained Molecular Dynamics Simulations of Organic Friction Modifier Adsorption on Rough Surfaces under Shear

Jiahao Tang [1], William Woei Fong Chong [2,3] and Hedong Zhang [1,*]

1 Department of Complex System Science, Graduate School of Informatics, Nagoya University, Furo-cho, Chikusa-ku, Nagoya 464-8601, Japan
2 Faculty of Mechanical Engineering, University Teknologi Malaysia (UTM), Johor Bahru 81310, Malaysia
3 Future Value Creation Research Center (FV-CRC), Graduate School of Informatics, Nagoya University, Furo-cho, Chikusa-ku, Nagoya 464-8601, Japan
* Correspondence: zhang@i.nagoya-u.ac.jp; Tel.: +81-052-789-4803

Abstract: Reducing friction energy losses is crucial in mechanical systems, often achieved through lubrication strategies employing friction modifiers. These additives adsorb onto surfaces, forming boundary film to prevent solid–solid contacts. However, atomistic simulation techniques used to study these additives often ignore surface roughness due to high computational cost. This study addresses this gap by employing Coarse-Grained Molecular Dynamics (CG MD) to investigate the impact of surface roughness on the adsorption of Organic Friction Modifiers (OFMs) under shear. Traditional self-diffusion methods prove inadequate for determining the damping coefficients in CG models because of strong OFM adsorption effects. Therefore, shear-induced motion is introduced for the coefficient determination. The simulation reveals that a symmetrical model (identical opposing surfaces) shows OFM slip, desorption, and re-adsorption trends on rough surfaces, while an asymmetrical model (smooth cylinder on a flat surface) demonstrates increased adsorption on rough flat surfaces (up to 60.9%) compared to smooth flat surfaces under similar shearing conditions. However, rough flat surfaces with a smaller wavelength (6 nm) exhibit faster OFM desorption along the asperity region, up to four times more than a 24 nm wavelength surface. This research emphasizes the importance of considering surface roughness in simulating OFM behavior for lubrication applications.

Keywords: coarse-grained; molecular dynamics; surface roughness; friction modifiers

1. Introduction

Enhancing lubricity for mating surfaces is imperative in mechanical systems to minimize their energy losses due to friction. This is often achieved by incorporating small quantities of friction modifier additives into low-viscosity base oils [1–5]. In contrast to commonly used friction modifiers that contain sulfur and phosphorus, such as Molybdenum Dithiocarbamate (MoDTC) and Molybdenum Dithiophosphate (MoDTP), Organic Friction Modifiers (OFMs) are composed solely of carbon, hydrogen, oxygen, and nitrogen, thereby garnering significant attention and usage in pursuing greener lubrication [2,6,7]. However, OFMs are generally not as effective as molybdenum compounds in reducing friction [8,9]. Therefore, it is crucial to further improve the performance of OFMs.

Typical OFM molecules comprise at least one polar head group and one alkyl chain tail. Previous studies have suggested that OFMs form adsorbed films on sliding surfaces, preventing direct solid–solid contacts, thereby reducing friction in the boundary and mixed lubrication regime [6,10–12]. The chemistry of OFMs and solid surfaces determines their interaction strength. Additionally, the morphology of solid surfaces, including surface roughness and textures, determines the pressure distribution in the contact region. Consequently, these factors affect the structural and mechanical properties of OFM adsorption films under shear. A thorough understanding of these correlations is essential for improving the friction reduction performance of OFMs.

Several advanced experimental technologies have been employed to investigate the surface adsorption of OFMs. Atomic Force Microscopy (AFM) has been utilized to measure the morphology and frictional properties of adsorbed films [13–17]. Quartz Crystal Microbalance (QCM) has been adopted to assess not only adsorption mass but also adsorption kinetics [18–21]. Additionally, Neutron Reflectometry (NR) has been used to analyze the structure of adsorption films in more detail, including the number of layers and the thickness and density of each layer [22–24]. These experimental approaches significantly contribute to understanding the effect of OFM chemistry on surface adsorption, primarily under static conditions. Regarding the polar head groups, which govern the surface adsorption strength, studies have indicated that carboxyl and amine groups are more effective than alcohol, ester, and nitrile groups [2,6,20,25]. Additionally, increasing the number of polar groups can enhance adsorption strength through multi-site adsorption or chelate effect [20,26–29]. Studies have also shown that straight tails, such as saturated or trans-unsaturated alky chains, form more densely packed adsorption films than kink tails, such as cis-unsaturated alky chains, resulting in better friction-reducing performance [15,30]. Nevertheless, measuring adsorbed OFM films on rough surfaces becomes challenging due to their typical thickness of approximately 1 to 2 nm [14,15]. Therefore, there remains a scarcity of research on the correlation between surface roughness and the adsorption efficacy of OFMs. Furthermore, the limitations of experimental technologies become apparent in the lack of thorough analyses concerning the distribution and mechanisms of adsorption and desorption on rough surfaces, especially in the presence of shearing [31].

To address this gap, Molecular Dynamics (MD) simulations are often employed to fundamentally elucidate the adsorption of molecules like friction modifiers at the atomic level. Shi et al. [32] investigated the impact of molecular polarity and temperature on the physisorption behavior of OFMs. Ewen et al. [33] focused on the slip and friction behavior of alkanes on monolayer OFM films adsorbed on solid surfaces during shear motion. Chen et al. [34] recently analyzed the adsorption behavior of *N*-(2,2,6,6-tetramethyl-1-oxyl-4-piperidinyl)-dodecaneamide (TEMPO-based) OFM on iron oxide surfaces, revealing the mechanism behind its experimentally proven superior lubrication performance. Concerning surface roughness, Eder et al. [35] studied different molecular surface coverages and rough substrates, focusing on the Derjaguin offset. Ewen et al. [36] demonstrated that OFMs can strongly adsorb onto surfaces with random nanoscale roughness (0.2–0.8 nm RMS) to prevent direct contact with solids under high-pressure conditions. Their study highlighted the correlation between nanoscale roughness and stearic acid coverage, indicating lower friction forces in systems with higher coverage.

While the above studies contributed valuable insights, it is crucial to note that the most widely used atomistic simulation techniques have size limitations, where the surface roughness dimensions were either minimized or the system was completely immersed in a lubricating solution [37–39]. One significant concern with this approach is the potential oversight of the space for liquid molecule dispersion out of confinement during shearing. This assumption may impact the number of molecules available for adsorption within the confinement.

To address the limitation and simulate larger systems that better mimic realistic contact conditions, we propose using Coarse-Grained (CG) models to simulate the adsorption behavior of OFMs on rough surfaces. CG models allow for the simulation of larger systems with significant surface roughness. However, there remains a challenge in developing CG models that can accurately reproduce the static and dynamic characteristics of target systems [40,41]. Expanding on our previous research [42], we introduce an enhanced approach to replicate the dynamic characteristics of CG models against All-atom (AA) models. This improved methodology is applied to larger-scale rough surface structures, enabling the study of the motion and distribution of OFMs during shearing processes in a more representative manner.

2. Simulation Details

2.1. Simulation Method

In this study, MD simulations were performed using CG models and Dissipative Particle Dynamics (DPD) [43–45]. The equation governing the motion of CG beads is expressed as follows:

$$M_i \ddot{\boldsymbol{r}}_i = \boldsymbol{F}_i^{\mathrm{C}} + \boldsymbol{F}_i^{\mathrm{D}} + \boldsymbol{F}_i^{\mathrm{R}}, \tag{1}$$

Here, M_i and $\boldsymbol{r}_i$ represent the mass and position of CG bead i, and $\boldsymbol{F}_i^{\mathrm{C}}$, $\boldsymbol{F}_i^{\mathrm{D}}$, and $\boldsymbol{F}_i^{\mathrm{R}}$ denote the conservative, damping, and random forces exerted on CG bead i, respectively. $\boldsymbol{F}_i^{\mathrm{C}}$ is derived from the negative derivative of interactive potentials between CG beads (CG potentials), and $\boldsymbol{F}_i^{\mathrm{D}}$ and $\boldsymbol{F}_i^{\mathrm{R}}$ are calculated as follows:

$$\boldsymbol{F}_i^{\mathrm{D}} = \sum_j \boldsymbol{f}_{ij}^{\mathrm{D}} = \sum_j \begin{matrix} -\zeta^{\parallel} w_{\mathrm{D}}^{\parallel}(r_{ij}) (\boldsymbol{e}_{ij} \cdot \boldsymbol{v}_{ij}) \boldsymbol{e}_{ij} \\ -\zeta^{\perp} w_{\mathrm{D}}^{\perp}(r_{ij}) [\boldsymbol{v}_{ij} - (\boldsymbol{e}_{ij} \cdot \boldsymbol{v}_{ij}) \boldsymbol{e}_{ij}] \end{matrix}, \tag{2}$$

$$\boldsymbol{F}_i^{\mathrm{R}} = \sum_j \boldsymbol{f}_{ij}^{\mathrm{R}} = \sum_j \begin{matrix} \sigma^{\parallel} w_{\mathrm{R}}^{\parallel}(r_{ij}) (\boldsymbol{e}_{ij} \cdot \boldsymbol{\theta}_{ij}) \boldsymbol{e}_{ij} \\ +\sigma^{\perp} w_{\mathrm{R}}^{\perp}(r_{ij}) [\boldsymbol{\theta}_{ij} - (\boldsymbol{e}_{ij} \cdot \boldsymbol{\theta}_{ij}) \boldsymbol{e}_{ij}] \end{matrix}, \tag{3}$$

Here, $\boldsymbol{e}_{ij} = (\boldsymbol{r}_i - \boldsymbol{r}_j)/r_{ij}$ denotes the unit vector from bead j to i, $\boldsymbol{v}_{ij} = \boldsymbol{v}_i - \boldsymbol{v}_j$ represent the relative velocity of bead i to j, and $\boldsymbol{\theta}_{ij}$ is a time-varying three-dimensional vector with normally distributed random numbers as its elements. The superscripts $\parallel$ and $\perp$ indicate directions parallel and perpendicular to $\boldsymbol{e}_{ij}$, respectively. The parameters ζ (referred to as damping coefficient) and σ and weight functions w_{D} and w_{R} serve to adjust the strength of damping and random forces. To achieve the canonical ensemble, the following relationships must hold for both the parallel and perpendicular components:

$$\sigma^2 = 2k_{\mathrm{B}}T\zeta, \tag{4}$$

$$[w_{\mathrm{R}}(r_{ij})]^2 = w_{\mathrm{D}}(r_{ij}) = \left[\max\left(0, 1 - r_{ij}/r_{\mathrm{c}}^{\mathrm{DPD}}\right)\right]^2, \tag{5}$$

where k_{B} is the Boltzmann constant, T is the temperature, and $r_{\mathrm{c}}^{\mathrm{DPD}}$ is the cutoff distance for the damping and random forces. For simplicity, we used $\zeta^{\parallel} = \zeta^{\perp}$ and $w_{\mathrm{D}}^{\parallel}(r_{ij}) = w_{\mathrm{D}}^{\perp}(r_{ij})$ in this study.

The CG potentials and damping coefficient must be determined a priori. The former affects the static and dynamic properties of simulation systems, whereas the latter solely affects the dynamic properties of simulation systems. Therefore, this study initially derived the CG potentials by matching static properties obtained from AA simulations and then determined the damping coefficient by aligning with dynamic properties from AA simulations.

All AA and CG MD simulations were conducted using the Large-scale Atomic/Molecular Massively Parallel Simulator (LAMMPS) [46], specifically with an in-house extended DPD package for the CG simulations [45]. The simulation results were visualized using the Open Visualization Tool (OVITO) [47].

2.2. Materials and Coarse-Grained Models

Dodecane ($C_{12}H_{26}$), stearic acid ($C_{18}H_{36}O_2$), and hematite plates with (100) surface orientation were used as base oil, OFM additive, and solid surfaces, respectively. As illustrated in Figure 1, dodecane and stearic acid molecules were coarse-grained by mapping three consecutive carbon atoms and their bonding atoms into a single bead, with the interaction center defined as the center of mass of the constituent atoms. Thus, each dodecane and stearic acid were modeled as four and six beads, respectively, connected by springs. The bead containing the carboxyl group was polar, while the others were nonpolar. For simplicity, the constituent beads of dodecane and stearic acid are collectively referred to as "liquid beads" hereafter. The hematite plates were coarse-grained by arranging solid beads in a simple tetragonal lattice with lattice constants $a = b = 5.714$ Å and $c = 5.814$ Å (see Figure 1).

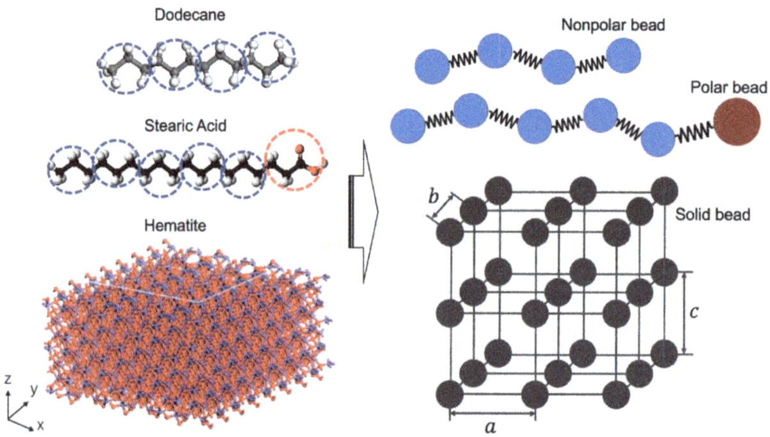

Figure 1. Schematic of coarse-grained models (**right**) mapped from all-atom models (**left**). a, b, and c are the lattice constants in the x, y, and z directions, respectively.

As depicted in Figure 2, CG potentials included bond–length potentials ($U_{\mathrm{bl-np}}$ and $U_{\mathrm{bl-p}}$), bond–angle potentials ($U_{\mathrm{ba-np}}$ and $U_{\mathrm{ba-p}}$), nonbonded liquid–liquid interaction potentials ($U_{\mathrm{ll-np}}$ and $U_{\mathrm{ll-p}}$), and nonbonded liquid–solid interaction potentials ($U_{\mathrm{ls}-np}$ and $U_{\mathrm{ls-p}}$). The subscripts "np" and "p" indicate involvement without and with polar beads, respectively, with $U_{\mathrm{ll-np}}$ being the only exception, which was also applied to polar and nonpolar liquid bead pairs. Notably, $U_{\mathrm{ll-p}}$ and $U_{\mathrm{ll-np}}$ were applied to liquid–bead pairs not involved in the bond–length and bond–angle interactions. These potentials were derived through the iterative Boltzmann inversion method, aligning with reference distribution functions characterizing static structural properties obtained from AA simulations, the same as in our previous studies [42,48]. Specifically, our previous work [42] extensively covered the derivation and validation of $U_{\mathrm{ll-np}}$, $U_{\mathrm{bl-np}}$, and $U_{\mathrm{ba-np}}$ for dodecane, demonstrating their high accuracy and transferability across a wide pressure range of up to 1 GPa. The remaining CG potentials, $U_{\mathrm{bl-p}}$, $U_{\mathrm{ba-p}}$, $U_{\mathrm{ll-p}}$, $U_{\mathrm{ls-p}}$, and $U_{\mathrm{ls-np}}$, alongside their validation, are detailed in Appendix A.

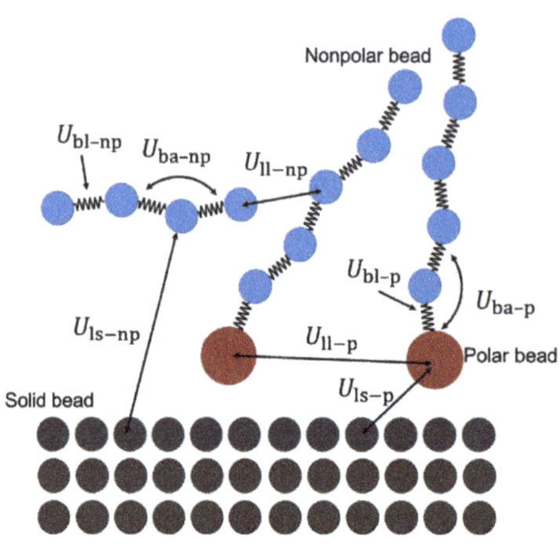

Figure 2. Schematic of all potentials between CG beads.

2.3. Determination of Damping Coefficients

The damping coefficients included ζ_{ll-p}, ζ_{ll-np}, ζ_{ls-p}, and ζ_{ls-np} for bead pairs of polar liquid, nonpolar liquid, polar liquid–solid, and nonpolar liquid–solid. Note that ζ_{ll-np} was also used for bead pairs of polar and nonpolar liquids. Conventionally, damping coefficients are adjusted by matching the self-diffusion coefficient of thermal motion obtained from AA simulations. However, due to the strong interactions of the polar end groups, stearic acid molecules form dimers in the bulk and adsorb on solid surfaces [49,50]. Therefore, thermally induced motion rarely occurs between the polar end groups and between them and solid surfaces, making it challenging to adjust ζ_{ll-p}, and ζ_{ls-p} by the conventional method, as evident from Equation (2). In this study, we addressed this challenge by employing shear-induced motion rather than thermal motion [51].

In brief, ζ_{ll-np} was initially determined by matching the shear viscosities from non-equilibrium AA and CG MD simulations of bulk dodecane systems under shear. Subsequently, ζ_{ll-p} was determined similarly using bulk stearic acid systems and the previously determined ζ_{ll-np}. For adjustment of ζ_{ls-np} and ζ_{ls-p}, we matched the Coefficients of Friction (CoF) from non-equilibrium AA and CG MD simulations, using a single layer of dodecane and stearic acid, respectively, lying flat between two hematite (100) surfaces, and applying a normal pressure and moving the upper solid surface horizontally. Appendix B provides details of the simulation setups and calculation of viscosities and CoF.

Figures 3 and 4 show the variation of bulk shear viscosities and CoF with the liquid–liquid and liquid–solid damping coefficients in the CG simulations, respectively. The values obtained from the parent AA simulations are also plotted in the figures with horizontal lines for comparison. These shear-related properties increase monotonically with the damping coefficients, showcasing their effectiveness in fine-tuning the damping coefficients. Table 1 summarizes the final values of ζ_{ll-np}, ζ_{ll-p}, ζ_{ls-np}, and ζ_{ls-p}, along with the corresponding shear-related properties from the CG and AA simulations. The good agreement validates the high accuracy of our CG models, even for reproducing dynamic properties.

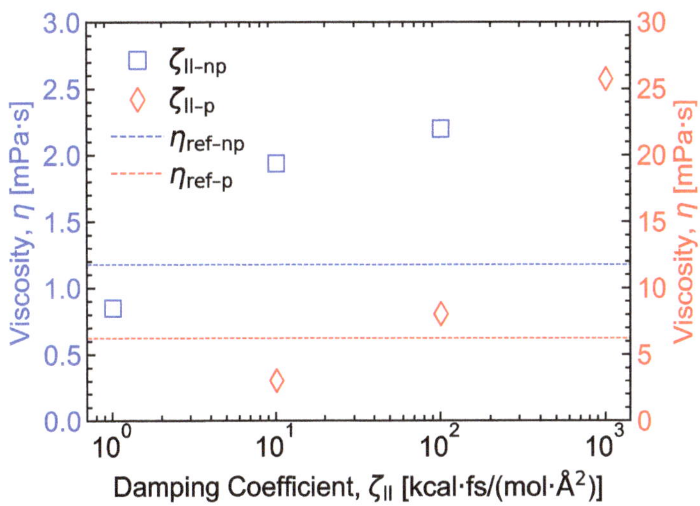

Figure 3. Variation of viscosity with the damping coefficients between liquid beads. Markers and horizontal lines indicate the results from CG and AA simulations, respectively.

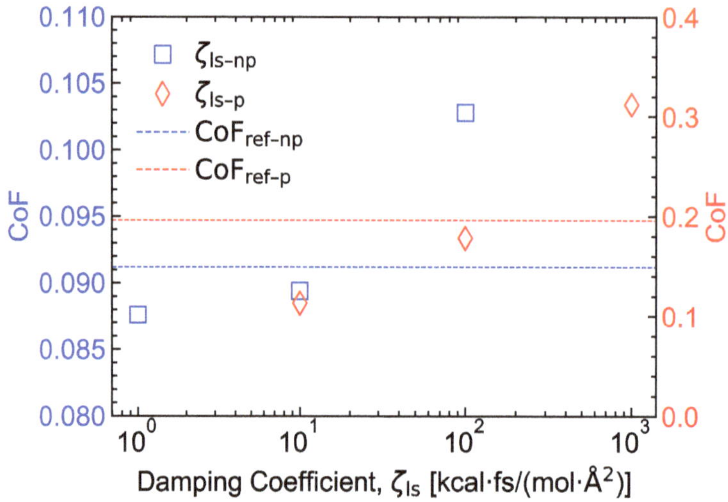

Figure 4. Variation of CoF with the damping coefficients between liquid and solid beads. Markers and horizontal lines indicate the results from CG and AA simulations, respectively.

Table 1. Determined values of damping coefficients and comparison of viscosity and CoF between coarse-grained and all-atom simulations.

ζ [(kcal·fs/(mol·Å^2)]	Coarse-Grained	All-Atom
	Viscosity, η [mPa·s]	
$\zeta_{\mathrm{ll-np}} = 4.1$	1.34	1.18
$\zeta_{\mathrm{ll-p}} = 36$	5.98	6.24
	CoF	
$\zeta_{\mathrm{ls-np}} = 18$	0.0906	0.0912
$\zeta_{\mathrm{ls-p}} = 132$	0.211	0.196

2.4. Models and Conditions of CG Shear Simulations

This study used two types of models, symmetric and asymmetric, for the CG shear simulations, as shown in Figures 5 and 6. The symmetric model had identical smooth or rough solid surfaces on the top and bottom, while the asymmetric model had a smooth cylindrical solid surface at the top and a smooth or rough flat surface at the bottom. In the symmetric model, the entire region between the solid surfaces was pressurized, whereas, in the asymmetric model, the pressure was only present in the region below the flat apex (24 nm in length) of the upper cylindrical surface, allowing liquid films to disperse from the pressurized region to surrounding spaces.

Both models had dimensions of 48 and 4 nm in the x and y directions, respectively, with periodic boundary conditions applied. A one-dimensional sinusoidal function $z = A\sin(2\pi x/\lambda)$ was used to describe the roughness of the upper and lower solid surfaces in the symmetric model and the lower solid surface in the asymmetric model. To capture the essence of surface roughness effect at reasonable computational costs, the amplitude A was fixed at 1.7 nm, and the wavelength λ was varied to 24, 48 nm, and infinity in the symmetric model, and 6, 12, 24, 48 nm, and infinity in the asymmetric model. It is to note that $\lambda = \infty$ means smooth surfaces. The radius of curvature of the cylindrical upper surface was 40 nm.

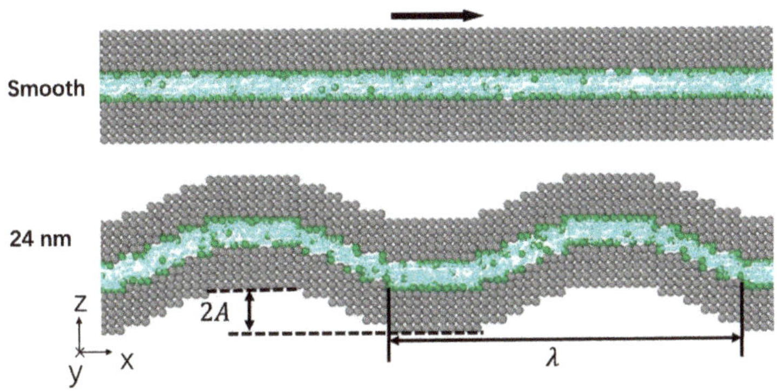

Figure 5. Front view of symmetric models. The gray, light blue, and green represent the solid, nonpolar liquid, and polar liquid beads, respectively. The top arrow indicates the shear direction of the upper solid.

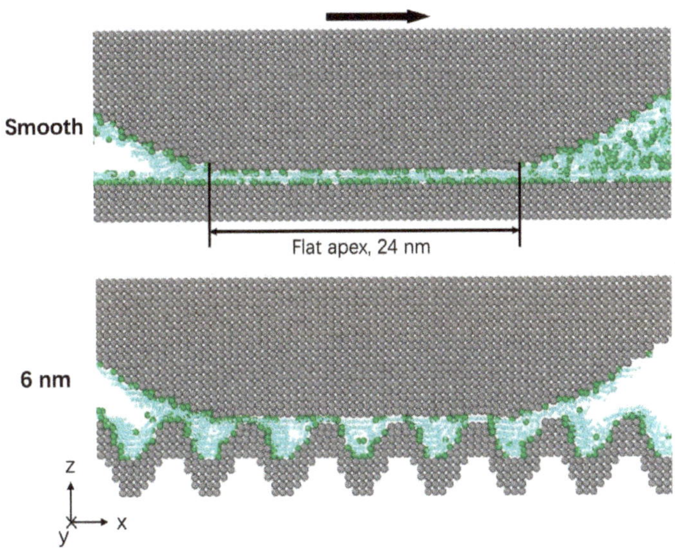

Figure 6. Front view of asymmetric models. The gray, light blue, and green represent the solid, nonpolar liquid, and polar liquid beads, respectively. The top arrow indicates the shear direction of the upper solid.

A monolayer physically adsorbed film was formed by placing the polar liquid beads of 168 stearic acid molecules on each solid surface, with the molecular chains perpendicular to the surface. This corresponded to an areal density of 0.875 molecules per nm^2. Then, a mixed solution of 300 dodecane and 300 stearic acid molecules was inserted into the central region between the two solid surfaces. The liquid films were then compressed by applying a constant pressure of 0.5 GPa on the upper solid surface for 0.5 ns. The shear simulation was performed by moving the upper solid surface at 10.0 m/s in the x direction while keeping the lower solid surface stationary. The system temperature was maintained at 353.15 K, and the simulation timestep was set to 5.0 fs. Movie S1 in the Supplementary Materials demonstrates the shear motion in the asymmetric model with a wavelength of 6 nm.

3. Results and Discussion

3.1. Symmetric Models

3.1.1. Adsorption Behavior

In the symmetrical model, Figure 7a depicts the time-dependent changes in OFM adsorption during the shearing of identical opposing surfaces. The adsorption criterion is that the distance between the OFM polar beads and their nearest solid beads is less than 0.7 nm, corresponding to the equilibrium distance of interaction between the polar liquid bead and the solid bead (U_{ls-p}). The adsorption fraction is calculated using $n(t)/N$, where $n(t)$ represents the number of adsorbed OFMs at time t, and $N = 636$ is the total number of OFMs. It is noted that the number of adsorbed OFM molecules presented is only along the bottom surface. Figure 7a reveals that the adsorption of OFM reaches a steady state after 20 ns; however, contrasting the stable saturation on the smooth surfaces, fluctuations in OFM adsorption are observed on the rough surfaces. The mean adsorption fraction is calculated from the final 30 ns of the simulations, as shown in Figure 7b. The adsorption fraction decreases slightly with increasing roughness wavelength, by 3.2% (24 nm) and 6.2% (48 nm) compared with the smooth surfaces. The error bars indicate that the variations in adsorption fraction on the rough surfaces are approximately twice as large as those on the smooth surfaces, emphasizing higher fluctuations on rough surfaces. These findings suggest that, in comparison to the stable adsorption of OFMs on smooth surfaces, as also confirmed in previous simulation studies [7,32], the adsorption of OFMs on rough surfaces is less stable.

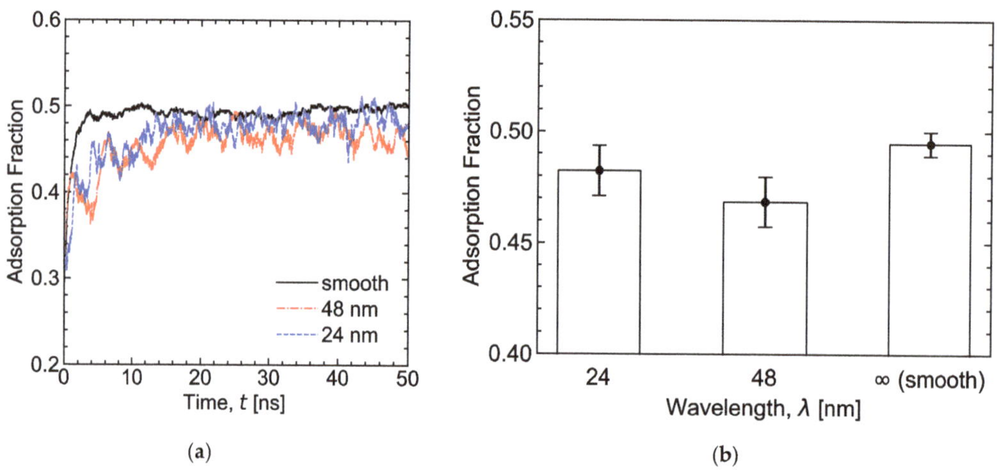

Figure 7. (**a**) Adsorption trend throughout the simulation duration in the symmetric model. (**b**) Mean and standard deviation (indicated by error bars) of adsorption fraction during the last 30 ns of the simulations for surfaces with different wavelengths.

3.1.2. Interpretation of Fluctuating OFM Adsorption Behavior

Figure 8a examines the fluctuating adsorption behavior of the OFM molecules on the rough surfaces in the symmetric model. Although the fluctuation period for the surfaces with a 48 nm wavelength is half that for the surfaces with a 24 nm wavelength, the reasons for the fluctuations are the same, as explained below. Therefore, the subsequent discussion will focus on the 24 nm wavelength. As observed at 18.8, 19.4, and 19.9 ns, there are three significant states of the fluctuations: local maximum adsorbed OFM molecules, transitioning state, and local minimum adsorbed OFM molecules. Snapshots of these states are provided in Figure 8b–d.

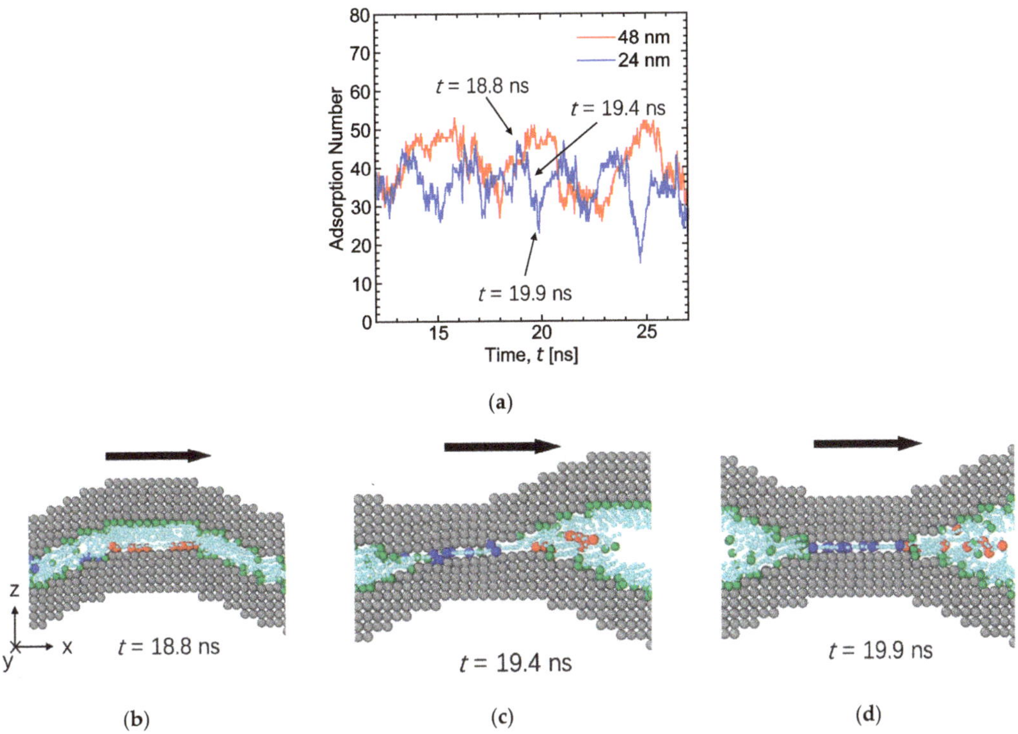

Figure 8. (**a**) Fluctuations in OFM adsorption on rough surfaces in the symmetric model. (**b–d**) Snapshots showing three characteristic states for adsorption of OFM molecules on the surfaces with 24 nm wavelength. Dodecane molecules are omitted for clarity. The gray, light blue, and green represent the solid, nonpolar liquid, and polar liquid beads, respectively. The red and blue indicate polar liquid beads that initially adsorbed and re-adsorbed onto the convex region of the lower surface, respectively. The arrows at the top of the snapshots indicate the shear direction of the upper solids.

To comprehend the movement of OFMs during shearing, attention is focused on a cluster of polar beads (depicted in red) already adhered to the convex areas of the lower corrugated rough surface in Figure 8b. Under sufficiently high pressure during shearing, this cluster of OFMs (in red) slips from the convex region as opposing peaks align, as in Figure 8c. As the shearing progresses, these OFM molecules slide away from the convex region, detaching and migrating to the concave region. Concurrently, another cluster of OFMs (depicted in blue) is sheared onto the vacant sites along the convex region. In Figure 8d, these OFM molecules (in blue) re-adsorb onto the convex region of the opposing surface. Meanwhile, the disengaged molecules (in red) are either carried away until they coincide with another convex region or re-adsorb along the transitional slope region of the surface feature. This slip, desorption, and re-adsorption cycle persists throughout the simulation, contributing to the fluctuating adsorption trend illustrated in Figure 7a. The cyclic behavior is attributed to the identical opposing rough surfaces (i.e., roughness amplitude and wavelength), potentially resulting in a less stable layer of adsorbed OFM.

3.2. Asymmetric Models

3.2.1. Adsorption Behavior

In the asymmetrical model, Figure 9a illustrates the variation in OFM adsorption over a 300 ns simulation as a smooth cylinder slides across a smooth, flat surface or a flat surface, with sinusoidal roughness wavelengths ranging from 6 to 48 nm. On the smooth surface, the plot indicates an initial increase followed by a gradual decrease in adsorbed OFM, finally leveling off into a steady state. This initial rise is attributed to the wedge effect induced by the cylindrical indenter, drawing in additional molecules, including those not initially adsorbed. Some of these molecules may adhere to available sites, increasing adsorption coverage within the confinement. However, sufficiently high pressure and shear forces could also lead to the slip and subsequent removal of adsorbed OFMs within the confined space. This cyclic process continues until a steady state is reached, typically around 200 ns in the simulation.

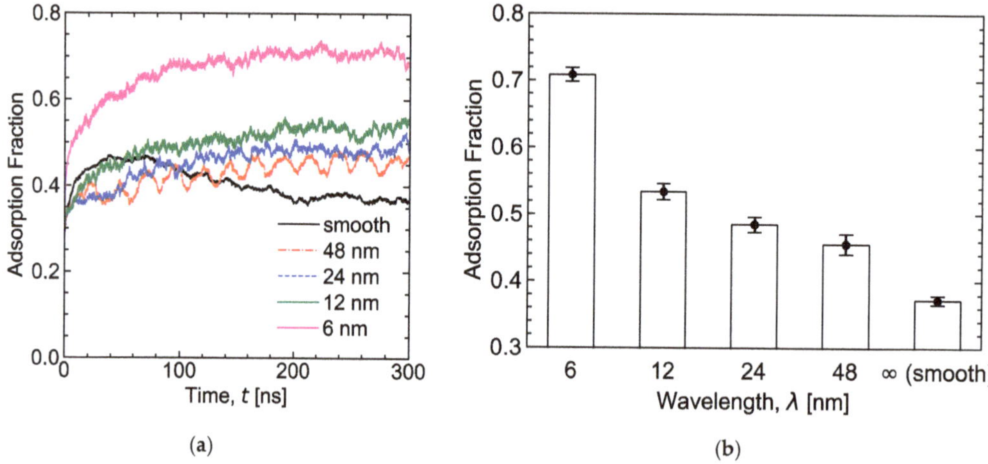

Figure 9. (**a**) Adsorption trend throughout the simulation duration in the asymmetric model. (**b**) Mean and standard deviation (indicated by error bars) of adsorption fraction during the last 100 ns of the simulations for surfaces with different wavelengths.

Similarly, simulations involving flat bottom surfaces with varying roughness reach a steady state at around 200 ns, displaying a consistent increase before saturation. The total number of adsorbed OFM molecules increases with smaller roughness wavelength values. Comparing this behavior with sliding on the smooth flat surface, the contact experiences an increase in mean OFM adsorption sites ranging between 9.9% (48 nm) and 60.9% (6 nm), as depicted in Figure 9b. The impact of surface roughness in the asymmetric model is clearly more pronounced than in the symmetric model shown in Figure 7b. Notably, the curvature of the indenter closely resembles the surface asperity curvature with a 48 nm wavelength, leading to a similar fluctuation as observed in Figure 7a. This behavior reflects the cyclic slip, desorption, and re-adsorption pattern of OFM molecules observed previously. The fluctuations diminish when the curvature of the indenter deviates from the surface asperity curvature.

The study further investigates specific locations on the rough surface where OFM molecules adhere, including the concave, convex, and transitional slope regions within the confinement, as shown in Figure 10. For all the rough surfaces, the adsorption fraction is the highest on the slope region. This is because that the total area of slope regions is the largest, providing more adsorption sites than the convex and concave regions. Moreover, due to the corner effect, more solid beads are in close contact with the polar beads adsorbed on the slope regions compared to the other regions, thereby resulting in stronger adsorption in the slope regions.

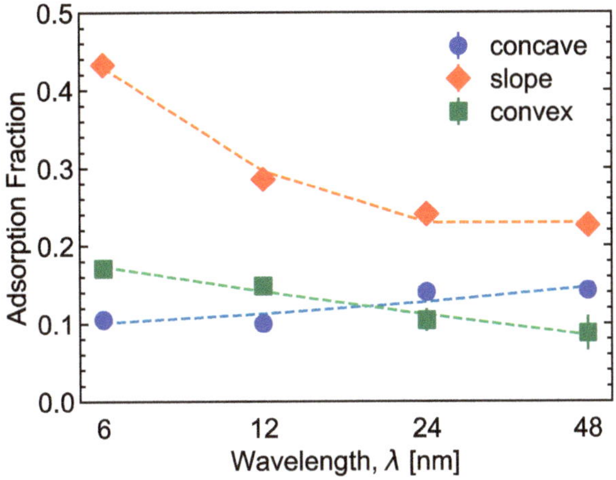

Figure 10. Average adsorption fraction over the last 100 ns in different regions of rough surfaces with varying wavelengths.

With decreasing wavelength, i.e., increasing surface roughness, the adsorption fraction increases on the slope and convex regions but decreases on the concave regions. Particularly, the change induced by surface roughness is more pronounced in the slope regions than in the convex and concave regions, which dictates the trend of increasing adsorption fraction with decreasing wavelength as observed in Figure 9b. The reasons are discussed below, based on the motions of OFMs as shown in Movie S2 in the Supplementary Materials. Due to the shear motion of the upper surface in the horizontal direction, OFMs possess horizontal rather than vertical velocity. As seen in Movie S2, OFM molecules not adsorbed onto the convex regions tend to traverse horizontally in the shallower cavity regions instead of diffusing vertically into the deeper cavity regions. If there are unoccupied slopes in the shallower regions, OFMs are likely to adsorb onto these slopes; otherwise, they move to the next convex region. As the wavelength decreases, although the total area of the slopes remains nearly constant, the area of slopes in the shallower region increases. This provides more favorable adsorption sites, resulting in a higher number of adsorbed OFMs on the slope regions. Additionally, with decreasing wavelength, the distance and travelling time of OFMs between neighboring convexes decrease. The shortened time make it more difficult for OFMs to diffuse to the deeper cavity regions, thereby leading to a decreased adsorption in the concave regions. Moreover, compared to OFMs in larger cavities of surfaces with longer wavelengths, OFMs in smaller cavities of surfaces with shorter wavelengths are more likely to entrain into the contacting interface, consequently resulting in an increased adsorption fraction in the convex regions.

3.2.2. Interpretation of OFM Desorption in Shearing

Focusing on the convex region, where boundary friction is expected to occur, the desorption time (t_d) for OFMs on a convex region is calculated as $t_d = t_{end} - t_0$, where t_{end} is the time at which the last initially adsorbed OFM leaves the original convex region, and t_0 is the time at which shearing commenced. Figure 11 presents the desorption time, indicating the duration needed for OFM, initially pre-adsorbed in convex regions, to completely desorb from their original locations. The plot illustrates a noticeable increase in the desorption time with larger roughness wavelength values. The variations in desorption time can be attributed to the increased number of surface peaks or asperities, with a smaller wavelength value correlating to an expedited OFM desorption process.

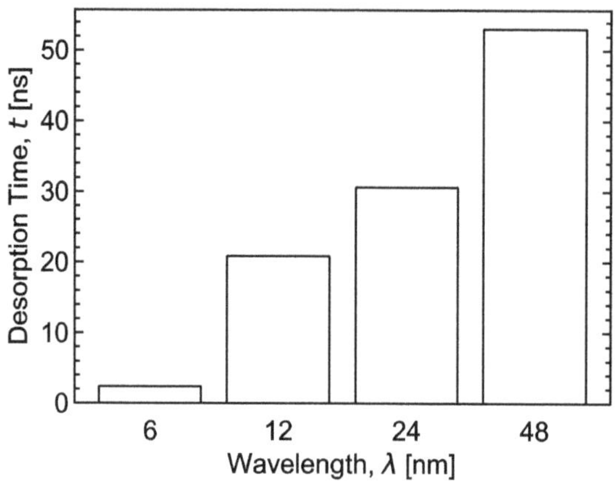

Figure 11. Time for OFM desorption from convex regions on rough surfaces with varying wavelengths.

To further interpret OFM desorption time, the current study employs the Greenwood and Williamson (GW) rough surface contact model [52]. The model assumes only one rough surface, similar to the asymmetric model simulated in the present study. Following this model, predictions for asperity contact area and contact load for the asymmetrical model, between the flat apex of the upper cylindrical surface and the rough flat surface, are made based on the following equations [52]:

$$\text{Total asperity contact area}\left(\text{m}^2\right) : A_a = \pi(\xi\beta\sigma)F_1(h/\sigma)A, \tag{6}$$

$$\text{Total asperity contact load (N)} : P_a = \frac{4}{3}\left(\xi\beta^{1/2}\sigma^{3/2}\right)F_{3/2}(h/\sigma)EA, \tag{7}$$

where E, A, and h refers to the reduced modulus of elasticity, apparent contact area, and gap between the upper and lower surfaces, in the units of Pa, m^2, and m, respectively. Equations (6) and (7) highlight that the primary roughness geometrical factors influencing asperity area and load are $(\xi\beta\sigma)$, and $\left(\xi\beta^{1/2}\sigma^{3/2}\right)$, respectively.

Utilizing the model, the rough surfaces generated with different wavelength values are translated into GW parameters, as detailed in Table 2. Since the amplitude remains constant, the variations in the adopted configurations of rough surfaces result in the same RMS roughness of 1.204 nm. The distinction lies in the mean asperity peak radius (β) and density (ξ). Following the GW parameters, a surface with a smaller wavelength value yields higher asperity density and smaller-sized asperities (based on curvature radius).

Table 2. Surface properties based on Greenwood and Williamson's rough surface contact model.

Wavelength [nm]	RMS Roughness, σ [nm]	Mean Asperity Peak Radius, β [nm] [53]	Mean Asperity Peak Density, ξ [nm^{-2}] [53]	Mean Asperity Slope (rad) [54]	$(\xi\beta\sigma)$ $(\times 10^{-2})$	$\left(\xi\beta^{1/2}\sigma^{3/2}\right)$ $(\times 10^{-3})$
6		0.5431	27.7778	1.1304	1.8170	27.0585
12	1.204	2.1477	6.9444	0.5654	1.7964	13.4524
24		8.5834	1.7361	0.2827	1.7948	6.7232

Table 2 reveals a consistent total asperity contact area across the various generated rough surfaces, while the load supported by surface asperities exhibits a significant difference. Surfaces with a 6 nm wavelength value are expected to carry an asperity load up to three times larger than those with a 24 nm wavelength value. The faster desorption rate observed for smaller wavelength values in Figure 11 is likely attributed to the increased load on surface asperities or the convex region, resulting in quicker desorption of OFM molecules. The faster desorption could also encourage more OFM molecules to migrate to the transitional slope region during shear, further increasing the adsorbed molecules in this region, as shown in Figure 10.

Additionally, Table 2 highlights the association between smaller wavelength values and higher mean asperity slopes. Sliding on such slopes often aligns with Euler's interlocking mechanism, where shear is proportionate to the tangent of the asperity slope. In scenarios involving potential solid wear particles (of similar length scale as the surface roughness) in the concave region, the surplus of adsorbed OFM molecules along the transitional slope between the concave and convex regions, particularly with smaller wavelength values, offers a more convenient pathway for these particles to slide out of the concave region with greater ease. Consequently, an enhanced number of adsorbed OFMs in this region is a mitigating factor, reducing the potential for erosive wear.

4. Conclusions

In this study, CG MD was employed to model OFM adsorption under shear in two distinct configurations: (1) symmetric model (identical opposing surfaces) and (2) asymmetric model (smooth cylinder on flat surface). The construction of the CG model faced challenges due to the strong effect of OFM adsorption and dimer cluster, making the traditional self-diffusion methods unsuitable for determining the damping coefficients of DPD. To address this challenge, shear-induced motion was employed, allowing the CG model to accurately match the viscosity of the AA model in liquid–liquid situations and the CoF of the AA model in liquid–solid situations. This approach yielded a more appropriate determination of the damping coefficients, improving the relevance of the CG model in simulating OFM behavior under shear conditions.

The symmetric model, considering identical opposing surfaces in shear, reveals that, despite a higher number of adsorbed OFMs by up to 6.2%, a distinct cyclic slip, desorption, and re-adsorption process occurs when opposing surfaces share similar roughness wavelength and amplitude during shear. This cyclic behavior, absent in the atomically smooth flat-on-flat configuration, emphasizes the complex nature of OFM adsorption behavior under shear conditions in the presence of surface roughness.

Simulations with a more realistic contact scenario with the asymmetric model involving a smooth cylinder sliding on a rough flat surface also show an overall increase in adsorbed OFMs, reaching a substantial 60.9%. This increase is observed primarily in a rough flat surface with a 6 nm wavelength compared to sliding on a smooth flat surface. However, it is important to note that this configuration's surplus of adsorbed molecules mainly accumulates along the transitional slope between convex and concave regions. This rough flat surface also exhibits faster OFM desorption along the convex or asperity region, attributed to the larger load carried by the asperity along the convex regions, reaching up to four times compared to a rough surface with a 24 nm wavelength.

In summary, this study observed the varying OFM adsorption and desorption behavior under diverse shear conditions, particularly concerning surface roughness. Unlike observations on atomically smooth surfaces, the cyclic patterns of OFM desorption and re-adsorption between rough surfaces, alongside enhanced adsorption in realistic contact scenarios, underscore the pivotal role of surface roughness in dictating the dynamic behavior of OFM adsorption under shear.

The study highlights the potential ramifications of overlooking surface roughness when simulating or designing OFM molecular structures. Such oversights may compromise the reliability of friction-reducing performance. Striking a balance between simulation details and complexity while accurately modeling rough surfaces is essential. This ensures a comprehensive understanding of OFM interactions under varying shear conditions, facilitating reliable predictions and informed design choices for OFMs in lubrication applications.

Supplementary Materials: The following supplementary information can be downloaded at: https://www.mdpi.com/article/10.3390/lubricants12020030/s1, Movie S1: Shear motion in the asymmetric model with a wavelength of 6 nm. Movie S2: Representative motions of OFMs observed in the asymmetric model with a wavelength of 12 nm. The other OFM and dodecane molecules are omitted for clarity.

Author Contributions: Conceptualization, J.T., W.W.F.C. and H.Z.; methodology, J.T. and H.Z.; software, J.T.; validation, J.T.; formal analysis, J.T.; investigation, J.T., W.W.F.C. and H.Z.; resources, J.T. and H.Z.; data curation, J.T.; writing—original draft preparation, J.T. and W.W.F.C.; writing—review and editing, J.T., W.W.F.C. and H.Z.; visualization, J.T.; supervision, W.W.F.C. and H.Z.; funding acquisition, H.Z. All authors have read and agreed to the published version of the manuscript.

Funding: This work was supported in part by JSPS KAKENHI Grants (Nos. 21H01238 and 23H05448).

Data Availability Statement: The data presented in this study are available in the article.

Acknowledgments: J.T. acknowledges the financial support provided by the THERS Interdisciplinary Frontier Next Generation Researcher Program (No. RD191018).

Conflicts of Interest: The authors declare no conflicts of interest.

Appendix A. Coarse-Grained Potentials and Validation

Appendix A.1. Liquid–Liquid Potentials

The figures below present the bonded potentials for bond–length (U_{bl-p}) and bond–angle (U_{ba-p}), as well as the nonbonded potential (U_{ll-p}) for polar beads, alongside their corresponding distribution functions. In all the distribution functions, the coarse-grained (CG) results reasonably closely match the all-atom (AA) results, validating the accuracy of the corresponding CG potentials.

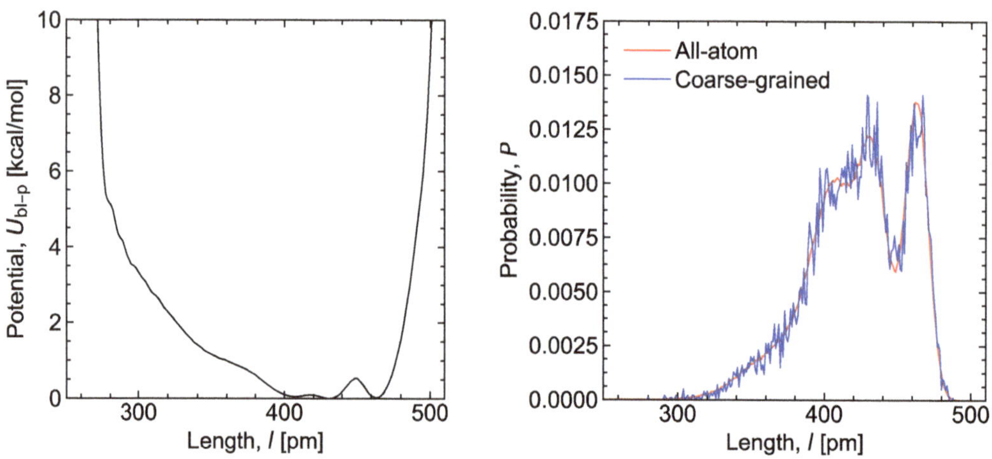

(**a**) (**b**)

Figure A1. (**a**) Coarse-grained bond-length potential involving polar bead, U_{bl-p}. (**b**) Bond length distribution functions derived from simulations of an isolated stearic acid using all-atom and our coarse-grained models.

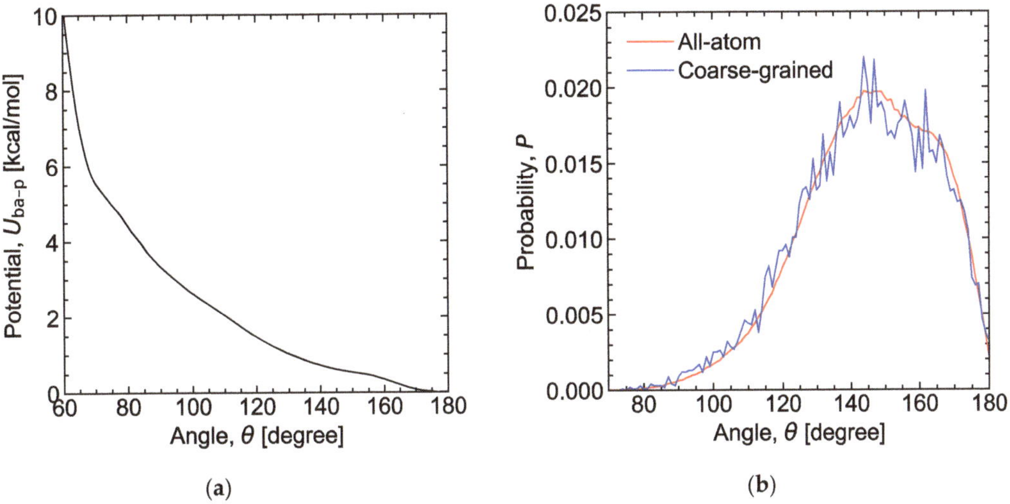

Figure A2. (**a**) Coarse-grained bond-angle potential involving polar bead, U_{ba-p}. (**b**) Bond angle distribution functions derived from simulations of an isolated stearic acid using all-atom and our coarse-grained models.

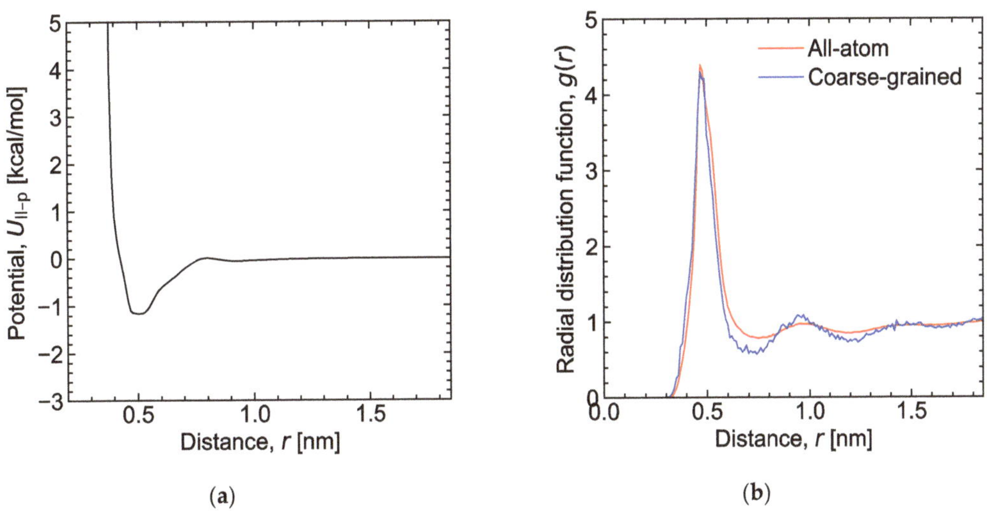

Figure A3. (**a**) Coarse-grained nonbonded polar liquid–liquid potential, U_{ll-p}. (**b**) Radial distribution functions of polar beads derived from simulation of bulk stearic acid systems using all-atom and our coarse-grained models.

Appendix A.2. Liquid–Solid Potentials

The figures below present the nonpolar and polar liquid–solid potentials, U_{ls-np} and U_{ls-p}, alongside the number density distributions (NDDs) of nonpolar and polar liquid beads versus their distance from the solid surface. In Figure A4b, the CG result strongly aligns with the AA one. Similarly, Figure A5b demonstrates a close match between the first peak in the CG and AA results. These validate the effectiveness of the CG liquid–solid potentials in reproducing the structures of dodecane and adsorbed stearic acid films on hematite surfaces.

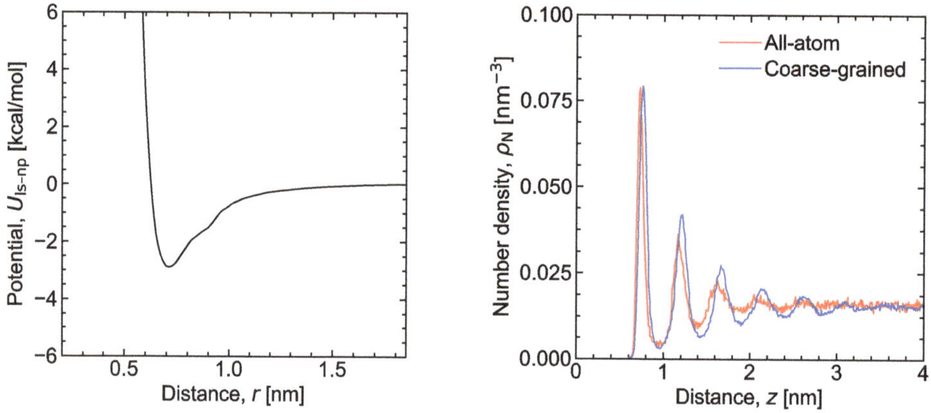

(**a**) (**b**)

Figure A4. (**a**) Coarse-grained nonbonded nonpolar liquid–solid potential, U_{ls-np}. (**b**) Number density distribution functions of nonpolar liquid beads derived from simulation of a dodecane film on a hematite plate using all-atom and our coarse-grained models.

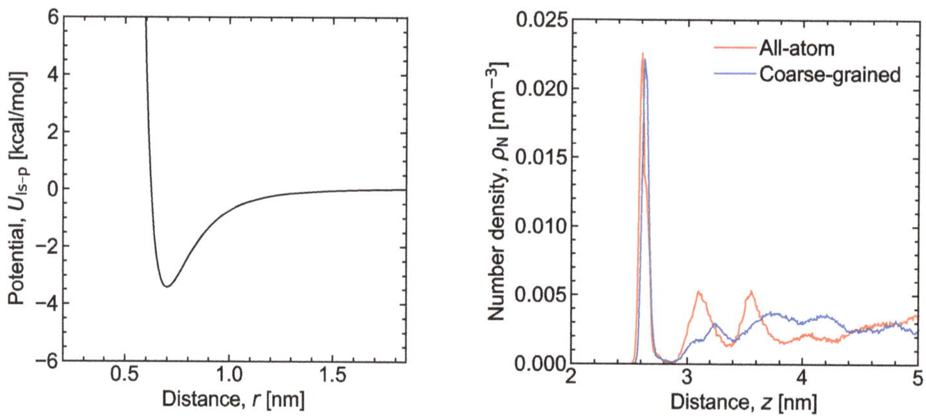

(**a**) (**b**)

Figure A5. (**a**) Coarse-grained nonbonded polar liquid–solid potential, U_{ls-p}. (**b**) Number density distribution functions of polar liquid beads derived from simulation of a stearic acid film on a hematite plate using all-atom and our coarse-grained models.

Appendix B. Simulation Setup for Fine-Tuning Damping Coefficients

Appendix B.1. Liquid–Liquid Damping Coefficients

As described in the main text, the liquid–liquid damping coefficients ζ_{ll-np} and ζ_{ll-p} were determined by matching the shear viscosities from non-equilibrium AA and CG simulations of bulk dodecane and stearic acid systems under shear, respectively. All simulations maintained a temperature of 353.15 K. The bulk dodecane system comprised 405 molecules in a box of 6.25 nm × 6.25 nm × 4.17 nm in the x, y, and z directions, yielding a density of 0.69 g/cm^3 at 0.1 MPa, which was the same as in our previous study [42]. The bulk stearic acid system contained 283 molecules in a box of 6.26 nm × 6.26 nm × 4.17 nm in the x, y, and z directions, producing a density of 0.83 g/cm^3, which approximates the experimental value under identical temperature and pressure conditions [55]. Periodic boundary conditions were applied to all the three directions.

To derive bulk viscosities, a constant engineering strain rate was applied to deform the simulation boxes and the motion of the systems was calculated using the SLLOD algorithm [51]. As depicted in Figure A6, shear was applied to the x–z plane in the x direction at a velocity of 1 m/s, generating a velocity gradient in the y direction. Viscosities were calculated using the equation $\eta = -P_{xy}/s$, where P_{xy} is the pressure tensor and $s\ (= v_x/l_y)$ is the shear rate with v_x being the shear velocity and l_y being the box length in the y direction. Each shear simulation spanned 3.1 ns. The individual viscosity values shown in Figure 3 and Table 1 in the main text are the average over the last 1 ns of three independent simulations.

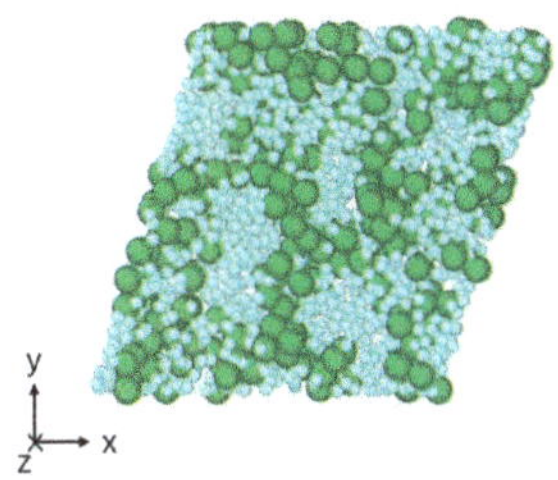

Figure A6. Snapshot from coarse-grained shear simulation of bulk stearic acid. The green and light blue represent the polar and nonpolar beads, respectively.

Appendix B.2. Liquid–Solid Damping Coefficients

As described in the main text, the liquid-solid damping coefficients ζ_{ls-np} and ζ_{ls-p} were determined by matching the coefficients of friction (CoF) from non-equilibrium AA and CG simulations, which involved the shearing of a monolayer layer of dodecane and stearic acid, respectively, between two hematite substrates. As depicted in Figure A7, the solid substrates measured 4.03 nm and 4.12 nm in the x and y directions, respectively, with periodic boundary conditions applied. The confined liquid film comprised 50 dodecane or 30 stearic acid molecules.

A normal pressure of 0.5 GPa and a shear velocity of 10 m/s in the x direction were applied to the top layer of the upper substrate, while the bottom layer of the lower substrate remained fixed. CoFs were computed by dividing the shear stress in the x direction of the x-y plane by the normal pressure. Each shear simulation spanned 5 ns. The individual CoF values shown in Figure 4 and Table 1 in the main text are the average over the last 2 ns of three independent simulations.

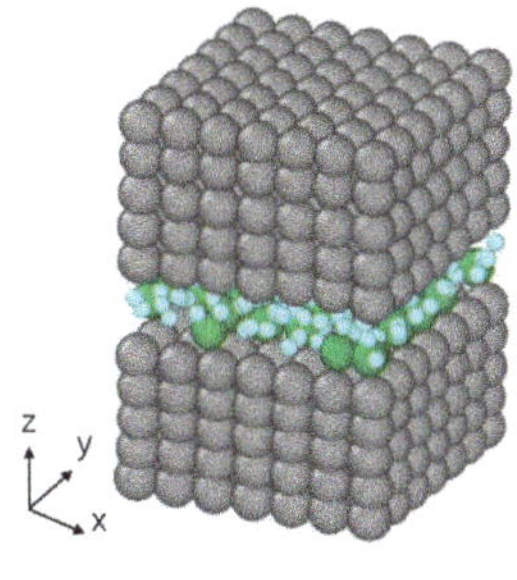

Figure A7. Snapshot from coarse-grained shear simulation of confined monolayer stearic acid film between solid substrates. The green, light blue, and gray represent the polar and nonpolar liquid beads and solid beads, respectively.

References

1. Holmberg, K.; Erdemir, A. Influence of Tribology on Global Energy Consumption, Costs and Emissions. *Friction* **2017**, *5*, 263–284. [CrossRef]
2. Spikes, H. Friction Modifier Additives. *Tribol. Lett.* **2015**, *60*, 5. [CrossRef]
3. McQueen, J.S.; Gao, H.; Black, E.D.; Gangopadhyay, A.K.; Jensen, R.K. Friction and Wear of Tribofilms Formed by Zinc Dialkyl Dithiophosphate Antiwear Additive in Low Viscosity Engine Oils. *Tribol. Int.* **2005**, *38*, 289–297. [CrossRef]
4. Zhang, J.; Meng, Y. Boundary Lubrication by Adsorption Film. *Friction* **2015**, *3*, 115–147. [CrossRef]
5. Vaitkunaite, G.; Espejo, C.; Wang, C.; Thiébaut, B.; Charrin, C.; Neville, A.; Morina, A. MoS2 Tribofilm Distribution from Low Viscosity Lubricants and Its Effect on Friction. *Tribol. Int.* **2020**, *151*, 106531. [CrossRef]
6. Cyriac, F.; Tee, X.Y.; Poornachary, S.K.; Chow, P.S. Influence of Structural Factors on the Tribological Performance of Organic Friction Modifiers. *Friction* **2021**, *9*, 380–400. [CrossRef]
7. Ewen, J.P.; Gattinoni, C.; Morgan, N.; Spikes, H.A.; Dini, D. Nonequilibrium Molecular Dynamics Simulations of Organic Friction Modifiers Adsorbed on Iron Oxide Surfaces. *Langmuir* **2016**, *32*, 4450–4463. [CrossRef]
8. Wang, W.; Li, C.; Yang, J.; Shen, Y.; Xu, J. Friction Performance of MoDTP and Ester-containing Lubricants between CKS Piston Ring and Cast Iron Cylinder Liner. *Lubr. Sci.* **2018**, *30*, 33–43. [CrossRef]
9. Cyriac, F.; Yi, T.X.; Poornachary, S.K.; Chow, P.S. Boundary Lubrication Performance of Polymeric and Organic Friction Modifiers in the Presence of an Anti-Wear Additive. *Tribol. Int.* **2022**, *165*, 107256. [CrossRef]
10. Tang, Z.; Li, S. A Review of Recent Developments of Friction Modifiers for Liquid Lubricants (2007–Present). *Curr. Opin. Solid State Mater. Sci.* **2014**, *18*, 119–139. [CrossRef]
11. Cañellas, G.; Emeric, A.; Combarros, M.; Navarro, A.; Beltran, L.; Vilaseca, M.; Vives, J. Tribological Performance of Esters, Friction Modifier and Antiwear Additives for Electric Vehicle Applications. *Lubricants* **2023**, *11*, 109. [CrossRef]
12. Shi, J.; Li, H.; Lu, Y.; Sun, L.; Xu, S.; Fan, X. Synergistic Lubrication of Organic Friction Modifiers in Boundary Lubrication Regime by Molecular Dynamics Simulations. *Appl. Surf. Sci.* **2023**, *623*, 157087. [CrossRef]
13. Hou, J.; Tsukamoto, M.; Zhang, H.; Fukuzawa, K.; Itoh, S.; Azuma, N. Characterization of Organic Friction Modifiers Using Lateral Force Microscopy and Eyring Activation Energy Model. *Tribol. Int.* **2023**, *178*, 108052. [CrossRef]
14. Campen, S.; Green, J.H.; Lamb, G.D.; Spikes, H.A. In Situ Study of Model Organic Friction Modifiers Using Liquid Cell AFM: Self-Assembly of Octadecylamine. *Tribol. Lett.* **2015**, *58*, 39. [CrossRef]
15. Campen, S.; Green, J.H.; Lamb, G.D.; Spikes, H.A. In Situ Study of Model Organic Friction Modifiers Using Liquid Cell AFM: Saturated and Mono-Unsaturated Carboxylic Acids. *Tribol. Lett.* **2015**, *57*, 18. [CrossRef]
16. Sahoo, R.R.; Biswas, S.K. Frictional Response of Fatty Acids on Steel. *J. Colloid Interf. Sci.* **2009**, *333*, 707–718. [CrossRef]
17. Hamdan, S.H.; Lee, C.T.; Lee, M.B.; Chong, W.W.F.; Chong, C.T.; Sanip, S.M. Synergistic Nano-Tribological Interaction between Zinc Dialkyldithiophosphate (ZDDP) and Methyl Oleate for Biodiesel-Fueled Engines. *Friction* **2021**, *9*, 612–626. [CrossRef]
18. Zachariah, Z.; Nalam, P.C.; Ravindra, A.; Raju, A.; Mohanlal, A.; Wang, K.; Castillo, R.V.; Espinosa-Marzal, R.M. Correlation Between the Adsorption and the Nanotribological Performance of Fatty Acid-Based Organic Friction Modifiers on Stainless Steel. *Tribol. Lett.* **2020**, *68*, 11. [CrossRef]
19. Gmür, T.A.; Mandal, J.; Cayer-Barrioz, J.; Spencer, N.D. Towards a Polymer-Brush-Based Friction Modifier for Oil. *Tribol. Lett.* **2021**, *69*, 124. [CrossRef]
20. Nalam, P.C.; Pham, A.; Castillo, R.V.; Espinosa-Marzal, R.M. Adsorption Behavior and Nanotribology of Amine-Based Friction Modifiers on Steel Surfaces. *J. Phys. Chem. C* **2019**, *123*, 13672–13680. [CrossRef]
21. Fry, B.M.; Moody, G.; Spikes, H.A.; Wong, J.S.S. Adsorption of Organic Friction Modifier Additives. *Langmuir* **2020**, *36*, 1147–1155. [CrossRef] [PubMed]
22. Shen, W.; Hirayama, T.; Yamashita, N.; Adachi, M.; Oshio, T.; Tsuneoka, H.; Tagawa, K.; Yagishita, K.; Yamada, N.L. Relationship between Interfacial Adsorption of Additive Molecules and Reduction of Friction Coefficient in the Organic Friction Modifiers-ZDDP Combinations. *Tribol. Int.* **2022**, *167*, 107365. [CrossRef]
23. Cyriac, F.; Yamashita, N.; Hirayama, T.; Yi, T.X.; Poornachary, S.K.; Chow, P.S. Mechanistic Insights into the Effect of Structural Factors on Film Formation and Tribological Performance of Organic Friction Modifiers. *Tribol. Int.* **2021**, *164*, 107243. [CrossRef]
24. Yamashita, N.; Hirayama, T.; Hino, M.; Yamada, N.L. Neutron Reflectometry under High Shear in Narrow Gap for Tribology Study. *Sci. Rep.* **2023**, *13*, 18268. [CrossRef] [PubMed]
25. Zhang, X.; Tsukamoto, M.; Zhang, H.; Mitsuya, Y.; Itoh, S.; Fukuzawa, K. Experimental Study of Application of Molecules with a Cyclic Head Group Containing a Free Radical as Organic Friction Modifiers. *J. Adv. Mech. Des. Syst. Manuf.* **2020**, *14*, JAMDSM0044. [CrossRef]
26. Hou, J.; Tsukamoto, M.; Hor, S.; Chen, X.; Yang, J.; Zhang, H.; Koga, N.; Yasuda, K.; Fukuzawa, K.; Itoh, S.; et al. Molecules with a TEMPO-Based Head Group as High-Performance Organic Friction Modifiers. *Friction* **2023**, *11*, 316–332. [CrossRef]
27. Hu, W.; Xu, Y.; Zeng, X.; Li, J. Alkyl-Ethylene Amines as Effective Organic Friction Modifiers for the Boundary Lubrication Regime. *Langmuir* **2020**, *36*, 6716–6727. [CrossRef]
28. Desanker, M.; He, X.; Lu, J.; Liu, P.; Pickens, D.B.; Delferro, M.; Marks, T.J.; Chung, Y.-W.; Wang, Q.J. Alkyl-Cyclens as Effective Sulfur- and Phosphorus-Free Friction Modifiers for Boundary Lubrication. *ACS Appl. Mater. Inter.* **2017**, *9*, 9118–9125. [CrossRef]
29. Desanker, M.; He, X.; Lu, J.; Johnson, B.A.; Liu, Z.; Delferro, M.; Ren, N.; Lockwood, F.E.; Greco, A.; Erdemir, A.; et al. High-Performance Heterocyclic Friction Modifiers for Boundary Lubrication. *Tribol. Lett.* **2018**, *66*, 50. [CrossRef]

30. Ouyang, C.; Bai, P.; Wen, X.; Zhang, X.; Meng, Y.; Ma, L.; Tian, Y. Effects of Conformational Entropy on Antiwear Performances of Organic Friction Modifiers. *Tribol. Int.* **2021**, *156*, 106848. [CrossRef]
31. Jacobs, T.D.B.; Ryan, K.E.; Keating, P.L.; Grierson, D.S.; Lefever, J.A.; Turner, K.T.; Harrison, J.A.; Carpick, R.W. The Effect of Atomic-Scale Roughness on the Adhesion of Nanoscale Asperities: A Combined Simulation and Experimental Investigation. *Tribol. Lett.* **2013**, *50*, 81–93. [CrossRef]
32. Shi, J.; Zhou, Q.; Sun, K.; Liu, G.; Zhou, F. Understanding Adsorption Behaviors of Organic Friction Modifiers on Hydroxylated SiO_2 (001) Surfaces: Effects of Molecular Polarity and Temperature. *Langmuir* **2020**, *36*, 8543–8553. [CrossRef] [PubMed]
33. Ewen, J.P.; Kannam, S.K.; Todd, B.D.; Dini, D. Slip of Alkanes Confined between Surfactant Monolayers Adsorbed on Solid Surfaces. *Langmuir* **2018**, *34*, 3864–3873. [CrossRef] [PubMed]
34. Chen, X.; Yang, J.; Yasuda, K.; Koga, N.; Zhang, H. Adsorption Behavior of TEMPO-Based Organic Friction Modifiers during Sliding between Iron Oxide Surfaces: A Molecular Dynamics Study. *Langmuir* **2022**, *38*, 3170–3179. [CrossRef] [PubMed]
35. Eder, S.J.; Vernes, A.; Betz, G. On the Derjaguin Offset in Boundary-Lubricated Nanotribological Systems. *Langmuir* **2013**, *29*, 13760–13772. [CrossRef] [PubMed]
36. Ewen, J.P.; Echeverri Restrepo, S.; Morgan, N.; Dini, D. Nonequilibrium Molecular Dynamics Simulations of Stearic Acid Adsorbed on Iron Surfaces with Nanoscale Roughness. *Tribol. Int.* **2017**, *107*, 264–273. [CrossRef]
37. Gao, J.; Luedtke, W.D.; Landman, U. Structures, Solvation Forces and Shear of Molecular Films in a Rough Nano-Confinement. *Tribol. Lett.* **2000**, *9*, 3–13. [CrossRef]
38. Bhushan, B.; Israelachvili, J.N.; Landman, U. Nanotribology: Friction, Wear and Lubrication at the Atomic Scale. *Nature* **1995**, *374*, 607–616. [CrossRef]
39. Math, S.; Gao, J.; Landman, U. Interfacial Segregation, Structure, and Diffusion of *n* -Alkane Mixture Films Adsorbed on Smooth and Rough Gold Surfaces. *J. Phys. Chem. C* **2022**, *126*, 4209–4219. [CrossRef]
40. Zhang, H.; Fukuda, M.; Washizu, H.; Kinjo, T.; Yoshida, H.; Fukuzawa, K.; Itoh, S. Shear Thinning Behavior of Nanometer-Thick Perfluoropolyether Films Confined between Corrugated Solid Surfaces: A Coarse-Grained Molecular Dynamics Study. *Tribol. Int.* **2016**, *93*, 163–171. [CrossRef]
41. Wang, H.; Junghans, C.; Kremer, K. Comparative Atomistic and Coarse-Grained Study of Water: What Do We Lose by Coarse-Graining? *Eur. Phys. J. E* **2009**, *28*, 221–229. [CrossRef] [PubMed]
42. Tang, J.; Kobayashi, T.; Zhang, H.; Fukuzawa, K.; Itoh, S. Enhancing Pressure Consistency and Transferability of Structure-Based Coarse-Graining. *Phys. Chem. Chem. Phys.* **2023**, *25*, 2256–2264. [CrossRef] [PubMed]
43. Plimpton, S. Fast Parallel Algorithms for Short-Range Molecular Dynamics. *J. Comput. Phys.* **1995**, *117*, 1–19. [CrossRef]
44. Junghans, C.; Praprotnik, M.; Kremer, K. Transport Properties Controlled by a Thermostat: An Extended Dissipative Particle Dynamics Thermostat. *Soft Matter* **2008**, *4*, 156–161. [CrossRef] [PubMed]
45. Kobayashi, T.; Zhang, H.; Fukuzawa, K.; Itoh, S. Effect of Transverse Dissipative Particle Dynamics on Dynamic Properties of Nanometer-Thick Liquid Films on Solid Surfaces. *Mol. Simulat.* **2020**, *46*, 1281–1290. [CrossRef]
46. Thompson, A.P.; Aktulga, H.M.; Berger, R.; Bolintineanu, D.S.; Brown, W.M.; Crozier, P.S.; In 'T Veld, P.J.; Kohlmeyer, A.; Moore, S.G.; Nguyen, T.D.; et al. LAMMPS—A Flexible Simulation Tool for Particle-Based Materials Modeling at the Atomic, Meso, and Continuum Scales. *Comput. Phys. Commun.* **2022**, *271*, 108171. [CrossRef]
47. Stukowski, A. Visualization and Analysis of Atomistic Simulation Data with OVITO–the Open Visualization Tool. *Model. Simul. Mater. Sci. Eng.* **2010**, *18*, 015012. [CrossRef]
48. Fukuda, M.; Zhang, H.; Ishiguro, T.; Fukuzawa, K.; Itoh, S. Structure-Based Coarse-Graining for Inhomogeneous Liquid Polymer Systems. *J. Chem. Phys.* **2013**, *139*, 054901. [CrossRef]
49. Yamamoto, S.; Matsuda, H.; Kasahara, Y.; Iwahashi, M.; Takagi, T.; Baba, T.; Kanamori, T. Dynamic Molecular Behavior of Semi-Fluorinated Oleic, Elaidic and Stearic Acids in the Liquid State. *J. Oleo Sci.* **2012**, *61*, 649–657. [CrossRef]
50. Wheeler, D.H.; Potente, D.; Wittcoff, H. Adsorption of Dimer, Trimer, Stearic, Oleic, Linoleic, Nonanoic and Azelaic Acids on Ferric Oxide. *J. Am. Oil Chem. Soc.* **1971**, *48*, 125–128. [CrossRef]
51. Hess, B. Determining the Shear Viscosity of Model Liquids from Molecular Dynamics Simulations. *J. Chem. Phys.* **2002**, *116*, 209–217. [CrossRef]
52. Greenwood, J.A.; Williamson, J.B.P. Contact of Nominally Flat Surfaces. *Proc. R. Soc. A Math. Phys. Eng. Sci.* **1966**, *295*, 300–319. [CrossRef]
53. Tomanik, E.; Chacon, H.; Teixeira, G. A Simple Numerical Procedure to Calculate the Input Data of Greenwood-Williamson Model of Asperity Contact for Actual Engineering Surfaces. *Tribol. Ser.* **2003**, *41*, 205–215. [CrossRef]
54. Bhushan, B. Surface Roughness Analysis and Measurement Techniques. In *Modern Tribology Handbook, Two Volume Set*; CRC press: Boca Raton, FL, USA, 2000; pp. 79–150.
55. Fisher, C.H. *N*-fatty Acids: Comparison of Published Densities and Molar Volumes. *J. Am. Oil Chem. Soc.* **1995**, *72*, 681–685. [CrossRef]

lubricants

MDPI

Article

Predicting Wear under Boundary Lubrication: A Decisive Statistical Study

Bernd Goerlach [1], Walter Holweger [2], Lalita Kitirach [3,*] and Joerg Fliege [2]

[1] ASC Goerlach, Robert-Bosch-Straße 60/1, 72810 Gomaringen, Germany; bg@asc-goerlach.de
[2] Operational Research Group, School of Mathematical Sciences, University of Southampton, Southampton SO17 1BJ, UK; walter.holweger@t-online.de (W.H.); j.fliege@soton.ac.uk (J.F.)
[3] Business School, Faculty of Social Sciences, University of Southampton, Southampton SO17 1BJ, UK
* Correspondence: lk1n22@soton.ac.uk

Abstract: The forthcoming revolution in mobility and the use of lubricants to ensure ecological friendliness intensifies the pressure on tribology for predictors in new life cycles, mainly addressing wear. The current paper aims to obtain such predictors by studying how the wear processes that occur in a standard FE8 bearing test rig under thin film lubrication are conducted by the properties of the lubricant rather than simple viscosity parameters. Assuming that the activity of a lubricant with respect to the temperature, surface, and chemicals is a matter of its chemical potential, the results show that the nature of the base oil is a key parameter, apart from the chemical structure of the additives. Moreover, it becomes clear that chemical predictors are changing by altering the conditions they are exposed to. As an important result, the lubricant is effective in the prevention of wear if it has the capacity to uptake and transmit electrical charges due to its polarisability during a wear process.

Keywords: tribology; wear; thin films; wear prediction; lubricants; chemistry; statistics

1. Introduction

1.1. A Few Aspects

The discipline of how power converts in a system of mating components with relative motion is the central topic of tribology. In a general view, at least two solid components are moved relatively against each other under force (the power input) and in the presence of a lubricant as a soft material in a randomly defined environment (see Figure 1).

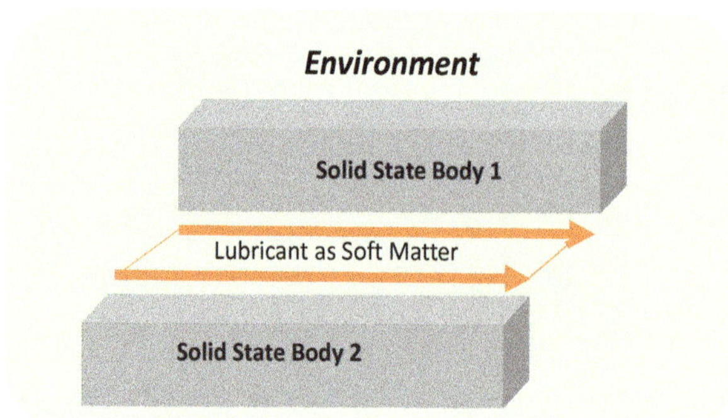

Figure 1. Abstract representation of a Tribosystem.

Tribology appears as a scattered cross-discipline, merging physical sciences and chemistry while adapting to self-generated theories. However, a concise definition is hard to state, and it is commonly established (see, for example, [1–5] and many more). The friction and lubrication of solids (solid-state bodies) were studied previously by Bowden and Tabor [6]. Lubrication as a soft matter medium, separating the solid-state bodies, has been extensively studied within EHL (elastohydrodynamic lubrication) (see, for example, [7–13] and many more). Basically, EHL is the topic of how a lubricant, as a body of soft matter, enters and passes a converging gap between two solid bodies. It undertakes the effort to define the role of the hydrostatic pressure that lifts off the moving solid parts. This effect reduces the direct touch of the solid bodies might lead to the consequence of wear. Hence, EHL is related to the Reynolds theory of fluid transportation in pipes (see, for example, [14–16] and many more). Similar to modern drive train technology (e.g., automotive), lubrication films narrow down in order to reduce the friction caused by splash losses, and thin films in the region of sub-microns and nanoscales appear. Thin film lubrication (TFL) became a field of intense research (see, for example, [17–20] and more). While within EHL and TFL, a lubricant is treated as a continuum, the continuous reduction in the tribofilm into a molecular dimension (a few nanometers) means that their chemical properties, e.g., electron transfer and photon emission processes, emerge. These processes are the domain of Tribo-catalysis (TC) (see, for example, [21] and more). Chemical processes related to catalysis are discussed as significant in understanding the role of functional additives in TFL and boundary lubrication. Plenty of studies have been carried out in order to investigate how the chemical structure is related to the function and the life of a drive train component (see, for example, [22] and more). Irregularities by the appearance of unexpectedly high wear and other effects caused by the use of additives that should prevent wear are reported (see, for example, [23,24] and more). Materials exposed to wear, where the solid bodies are touching each other, act under transient and non-equilibrium conditions (see, for example, [25–27] and more), emitting and exchanging photons, electrons, charged and non-charged particles (see, for example, [28–30] and more), denoted commonly as fracto-emission (FE). Lowering the film thickness in tribological contact brings both solid-state and lubricants under perpetual stress, causing the excitation of the molecular and solid-state structure. As the input frequency of stress is high compared to the relaxation and repairs the capability of the partner, electron and photon exchange processes appear, but also chemical reactions. The engineering of the transition from TFL to TC and FE is hard to parametrise by the components tested due to the many apparent scales (from a few nanometers to meters). Both engineering molecular processes and trans-scale modeling have given the motivation to undertake multi-scale modeling (see, for example, [17,31–33] and more).

1.2. White Etching Cracks (WEC) and Lubricants

WEC came into the interest of industry and research within more than a decade [34]. The failure has been attributed to material degradation processes (see, for example, [35,36] and more) in relation to lubricants and parametrised conditions (see, for example, [37,38] and more). Even though some patterns with respect to the compositions of lubricants are reported (see, for example, [39,40] and more), a clear pattern in relation to the structure of any chemical and stoichiometry is missing.

1.3. A Distinct Perspective

The continuous reduction in the tribogap (schematical sketch, see Figure 2) puts the lubricant constituents under stress. A (see Figure 2) reflects the situation where the thermal motion and mutual electronic excitation by neighbouring molecules are in balance with relaxation to the ground state. This basically reflects EHL and viscous flow. As their collision rates increase and the space becomes narrower, the molecules react with density and increase in symmetry, and their narrowing could lead to a superstructure (see Figure 2B). This self-organised structure acts as a new species, different in physics (e.g., A), solubility, viscosity, friction, and functional release. Narrowing the tribogap further (see Figure 2C) finally causes the disruption of the molecular system as the collision rates and the mutual touchdown do not allow structural relaxation.

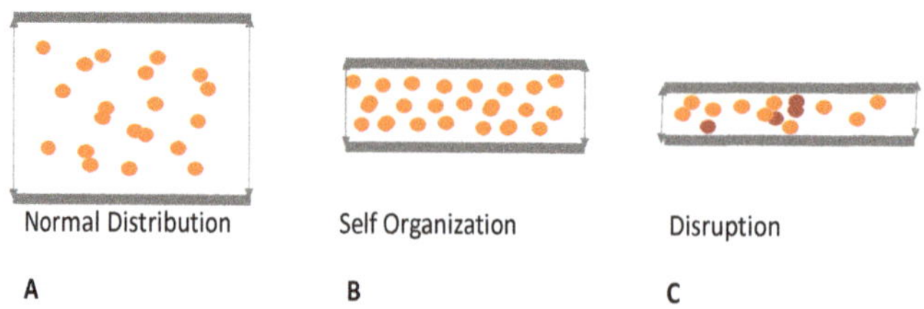

Figure 2. Reducing the tribogap from (**A**) (EHL) to (**B**) (TFL) to (**C**) (TC) as a scheme. Red dots shall mark cluster formation as consequence of spatial stress.

We argue that the capability of a lubricant to undergo the transition from A → B → C is a key parameter in lubrication.

2. Materials and Methods

With reference to the basic investigations, as described in the previous section, we tried to set up an empirical-based concept on how the molecular transition from EHL toward TFL and TC could be described. As lubricants in the life cycle are a) constituted of many components and b) exposed to complex loading, we took the concept presented earlier [41–43]. Basically, the mutual interaction therein is described as a matter of a non-dimensional factor *Di,j,* which arises from the chemical potential of a pair i,j of constituents. *Di,j* could be interpreted as a lubricant stress factor caused by impact stress from the tribological gap (see Figure 2). This concept neglects the equilibrium that each molecular system tries to reach under stress but describes instead the proximity or distance to the equilibrium. Ideally, the equilibrium is reached if *Di,j* = 1 is the chemical potential of a set of equal components i,j, which, per definition, means *Di,j* = 1. As an activity parameter, we took, according to [41–43], the dipolemoment divided by the sum of all dipolemoments as well as the induced dipolemonent divided by the sum of all induced dipolemoments as non-dimensional parameters to calculate the stress response factor, Dij, between the components i and j. If *Di,j* < 1, it stays within {0,1} and denotes it's probability to be present in the system. If, in contrast, *Di,j* > 1, we define the appearance of clusters by dividing *Di,j* by a number n (n ∈ Z), so that *Di,j* comes back into {0,1}. The number n is interpreted as degeneration, as the factor leading to clusters within the system of constituents. Notably, *Di,j* and the clusters appear simply as transient probabilities independent of how far they are from a thermodynamic equilibrium. Within a real lubricant, all combinations i,j have to be calculated by means of increasing activity. This leads to a matrix of cluster and non-cluster presence for each activity level on all possible combinations i,j related to the given activity. As all combinations are allowed per activity, the individual percentages of the dipole (cluster/non-cluster) and inducible dipoles (polarisabiliy) (cluster/non-cluster) are calculated. The dipole moments and polarisabilites are calculated using the semiempirical PM3 method (Hyperchen 8.0™). The chemical structures are derived either by brand name or by retro-analysing the lubricant with infrared and XRF analyses. All calculations described here are automised using the system software tool Solu42™ 1.1.

3. Experimental Section

In order to evaluate the system, an FE8 test rig (according) to the DIN 51819 standard was used (see Figures 3 and 4 (taken from [39]).

A set of test lubricants divided into Extreme Pressure (EP) and Anti-Wear (AW) candidates given by the chemical structure of the additives was used (see [43]) and Appendix A. All permutations for any component pair i,j as a function of a virtual activity were calculated using Software Package Solu42™. For the driving factor for the interaction of the components, their activity coefficient taking the relative dipolemoment was finally given by the dipolemoment of an individual divided by the sum of all components used. The same investigation was used to calculate the relative polarisabilities (the induced dipole moment). The dipolemoments and polarisabilities of all components are retrieved via a calculation using the semiempirical parametrised method 3 (PM3) with Hyperchem(8.0)™. As explicitly explained in [43], the factor Di,j is obtained as follows:

$$Di,j = \ln(ai)/\ln(aj) \quad (1)$$

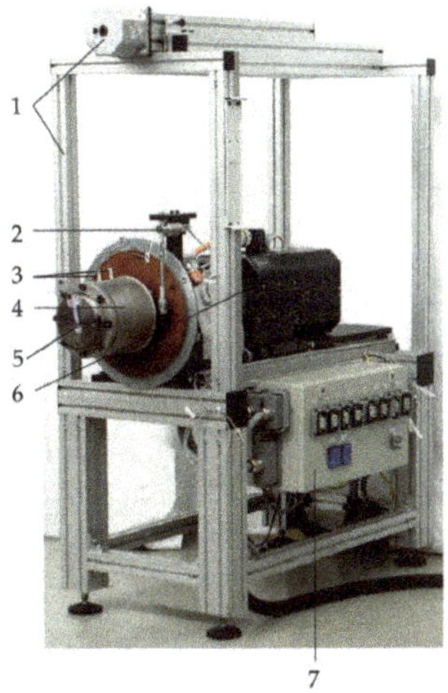

Figure 3. Structure of the FE8 test rig: 1. Frame and crane for the test head. 2. Torque transducer 3. Thermocouples for each bearing. 4. Test head. 5. Acceleration sensor 6. Driving unit 7. Control cabinet for terminals for cooling and fan heating. (Source [39]).

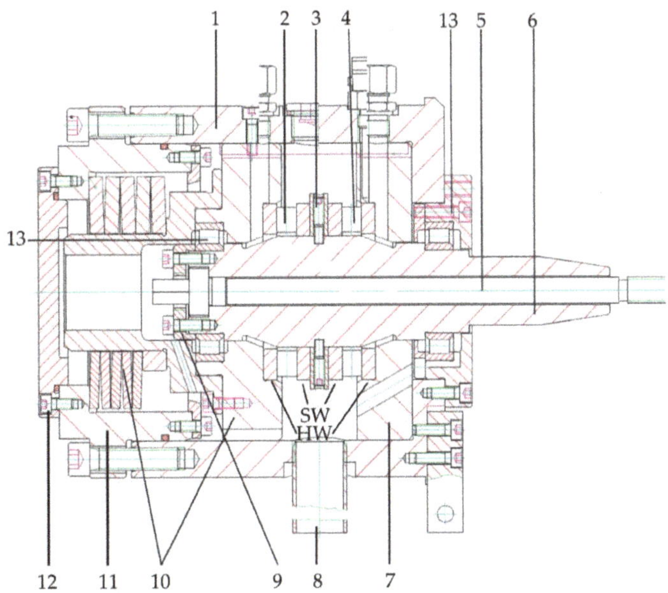

Figure 4. Test head with axial cylindrical roller bearings. 1. Housing. 2. Test bearing 2 (test head-sided bearing). 3. Spacer 4. Test bearing 1 (motor sided bearing). 5. Shaft. 6. Clamping bolt. 7. Bearing seat. 8. Drainpipe. 9. Cap. 10. Bearing support with screwed-on pilot pin. 11. Lid cup of spring package. 12. Lid. 13. Auxiliary bearing. Both test bearings consist of stationary housing (HW) and a rotating shaft washer (SW). Figure adapted from Source [39] and [43]).

(*ai*): Dipole activity for the relative dipolemoments and polarisability activity for the relative polarisability. The term *Di,j* is calculated using means in relation to increasing tribological activity (*aj*), which is basically the activity with respect to the surface (S) and the "inner" structure of the fluid (L).

If *Di,j* > 1, it is interpreted as a cluster (CL). Calculating all combinations, it comes down to the following predictors:

CL_Di_Add_S: Dipolar cluster formation of the additive at the surface
CL_Po_Add_S: Polarisability cluster formation of the additive at the surface
Di_Add_S: Dipolar formation of the additive at the surface
Po_Add_S: Polarisability formation of the additive at the surface
CL_Di_Add_L: Dipolar cluster formation of the additive within the liquid (L)
CL_Po_Add_L: Polarisability cluster formation of the additive within the liquid (L)
Di_Add_S_L: Dipolar formation of the additive within the liquid (L)
Po_Add_S_L: Polarisability formation of the additive within the liquid (L)
CL_Di_L_S: Dipolar cluster formation of the liquid at surface S
CL_Po_L_S: Polarisability cluster formation of the liquid at surface S

If more than one Baseoil component is present, e.g., Li,j, the combinations are extended for each i,j.

CL_Di_Add_Li,j: Dipolar cluster formation of the additive within the liquid Li,j
CL_Po_Add_Li,j: Polarisability cluster formation of the additive within the liquid Li,j
Di_Add_S Li,j: Dipolar formation of the additive within the liquid Li,j
Po_Add_S_ Li,j: Polarisability formation of the additive within the liquid Li,j
CL_Di_Li,j: Dipolar cluster formation of the additive within the liquid Li,j
CL_Po_Li,j: Polarisability Cluster formation of the Additive within the liquid Li,j
Di_Li,j_S: Dipolar formation of the liquid Li,j at surface S
Po_Li,j_S: Polarisability formation of the liquid Li,j at surface S
CL_Di_Li,j_S: Dipolar cluster formation of the liquid Li,j at surface S
CL_Po_Li,j_S: Polarisability cluster formation of the liquid Li,j at surface S

4. Results

The results of the component testing versus the variants in lubricants and their chemical predictors are exemplified in Table 1 for activity parameter 1, which represents the lowest activity. For the computation of the entirety of the chemical predictors, we used Hyperchem 8.0™ as a base in order to calculate the dipolemoment and the polarisability of each component. The computation of the chemical predictors as a function of imposed surface activity was carried out by the software package Solu42™. The results were split into non-clusters and clusters (CL) for the various combinations given by the base oil and the additives. For computation, an increase in surface activity was used (for details, please contact the authors). In order to obtain the main predictors explaining the individual wear of the test rig components, e.g., cage, rollers, and washers, but also the average wear and the wear of single components, a variety of statistical methods were used.

The boxplot is a standardised graphical representation that effectively illustrates the distribution of data by utilising its five-number summary, which includes the minimum value, the first quartile (Q1), the median, the third quartile (Q3), and the maximum value. Boxplots provide valuable insights into the presence of outliers and their corresponding values, as well as the symmetry of data, the degree of data clustering, and the skewness of data [44]. We calculated the average wear for each lubricant to better represent its corresponding wear. Figure 5 illustrates that the average wear of AW has outliers (2 dots in red rectangle shape). After removing these outliers, a clearer representation of the distribution of the average wear for both EP and AW is shown in Figure 6. The spread of average wear for EP is noticeably wider than AW. Despite their medians being closely matched, EP displays a right-skewed distribution, while AW leans toward a left-skewed pattern.

Table 1. Testing results (example from virtual activity 1).

Lubricant Code	Friction	Wear Components				Predictors									
		Wear Roller (mg.)	Wear Cage (mg.)	Wear Housing (mg.)	Wear Motorside (mg.)	Di_Add_S	CLDi_Add_S	Po_Add_S	CLPo_Add_S	Di_L1_S	CLDi_L1_S	Po_L1_S	Di_Add_L1	CLDi_Add_L1	CLPo_Add_L1
AW1	31.5	0.5	34	0.5	1	0	14.45	0	16.87	37.01	0	10.64	0	16.55	4.48
AW10	24.25	30.5	221	103.25	100.75	0	13.63	0	15.06	35.38	0	14.62	0	15.42	5.89
AW10a	57.25	2095	1030.75	1252.75	1402.25	0	31.59	0	11.98	0	14.57	12.48	23.21	0	6.17
AW11	18	131.5	138.25	141	140.25	0	15.79	0	16.9	29.21	0	16.92	0	12.18	9
AW11a	24.25	36.75	149.5	36.25	50.5	0	31.17	0	11.78	0	12.81	12.64	25.69	0	5.9
AW1a	33	0	356	0	0	0	33.39	0	13.37	0	15.32	9.12	24.66	0	4.15
AW6	31.5	68.5	95.75	162.25	168.25	0	14.07	0	15.91	35.5	0	12.62	0	16.33	5.57
AW6a	65.75	624.25	386	963.5	859.5	0	32.81	0	12.76	0	14.98	10.9	24.36	0	4.19
EP 14c	19.5	46	24	30.5	25.5	0	25.05	0	10.19	27.56	0	7.06	0	26.61	3.52
EP 14d	19.5	4.75	10.5	3.25	2.75	0	28.94	0	12.11	0	28.47	8.63	0	17.76	4.09
EP 1d	24	28.5	58.25	19	19.5	0	19.86	0	23.6	0	21.93	8.23	21.79	0	4.6
EP 2b	31.32	201.25	415.75	216	219.75	0	22.94	0	26.72	0	14.39	8.99	0	21.6	5.37
EP 6	24.8	122.75	265.25	153.75	150	0	29.41	0	29.26	13.96	0	14.43	0	6.59	6.35
EP 6a		9	378.25	8.5	4.75	0	22.35	0	25.16	0	9.63	13.24	21.01	0	8.6
EP 6e	22.5	109.75	275.25	108.25	134.25	0	18.05	0	17.97	21.57	0	22.3	0	8.23	11.88
EP 6f	26	1.75	255.25	3	1.25	0	29.52	0	10.87	0	10.59	14.39	27.84	0	6.79
EP1	22.41	73.5	387.5	111.5	112.75	0	26.63	0	26.74	14.56	0	14.19	0	8.81	9.08
EP14	20.06278259	137.25	249.25	170.75	172.5	0	30.93	0	31.18	12.38	0	11.8	0	8.04	5.67
EP14a	30.18226616	7	27.75	5.75	5.75	0	23.07	0	26.87	0	9.44	10.91	22.81	0	6.9
EP1a	29.60352298	97.25	53.5	57	62.25	0	22.44	0	25.04	0	10.09	14.15	20.23	0	8.05
EP1b	23.78	140.75	480.75	174.5	150.75	0	21.63	0	23.81	0	17.38	13.36	0	16.13	7.7
EP1c	18.5	94	149.5	119.25	99.75	0	22.97	0	26.81	0	15.2	9.11	0	20.57	5.34
EP1e	21	91.25	368.75	132	148.5	0	16.83	0	16.88	23.13	0	22.53	0	10.17	10.47
EP1f	25.25	138.75	57.5	69.25	89.75	0	29.61	0	10.82	0	11.08	15.37	26.77	0	6.35
EP2	20.53146015	171.75	383.75	212.5	186	0	33.15	0	33.46	11.1	0	10.53	0	6.69	5.08
EP2a	34.00044352	132.25	41.25	59	60.5	0	19.83	0	23.69	0	23.07	8.03	20.66	0	4.73

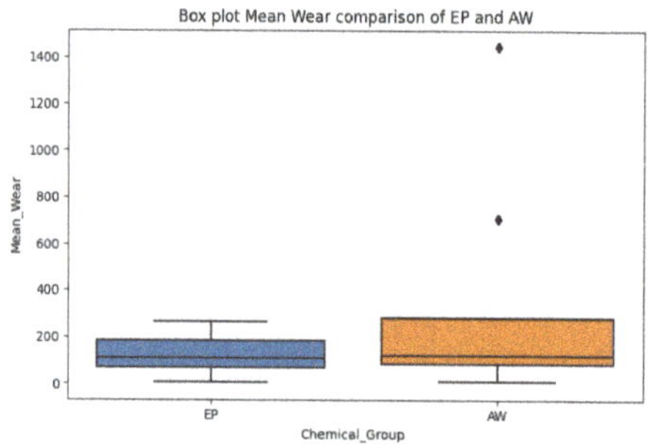

Figure 5. Box plot comparing the average wear between EP and AW (excluding outliers).

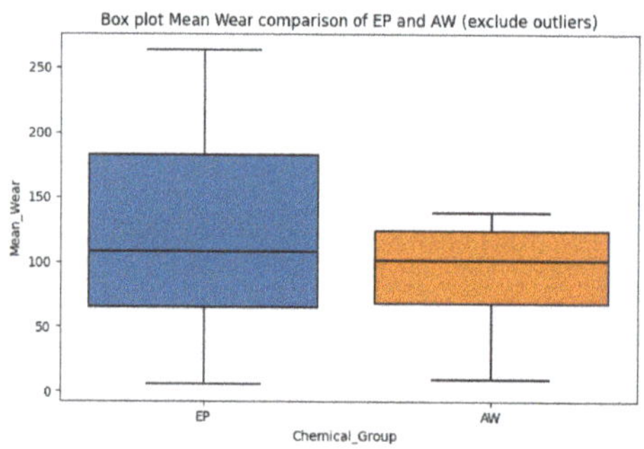

Figure 6. Box plot comparing the average wear between EP and AW.

4.1. Comparison of Wear Characteristics between EP and AW Lubricants

From Figure 7, overall, the median lines on the box plots suggest that EP lubricants tend to have higher wear attributes than AW across all four wear types. For the roller and cage types, AW data cluster tightly around the median and show much lower wear compared to EP. By contrast, for the housing and motorside types, the wear data for AW are more spread out, with the medians slightly different from EP.

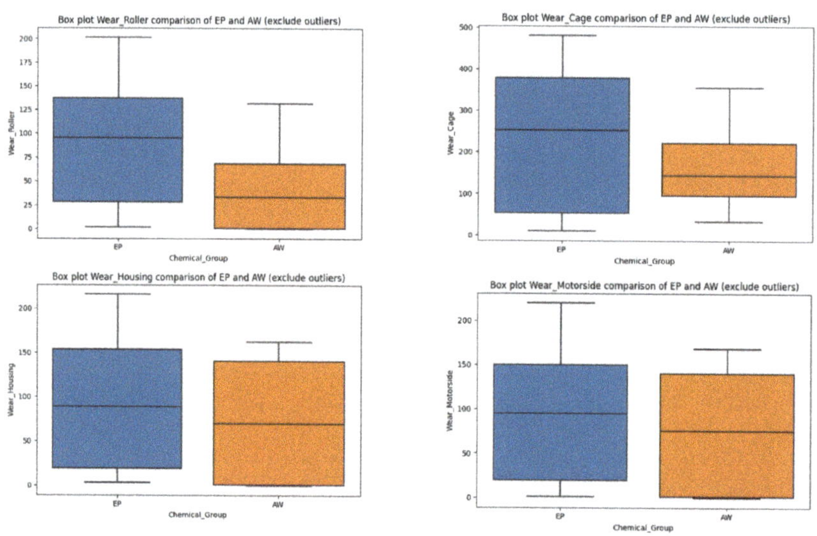

Figure 7. Box plot comparing the wear components between EP and AW.

4.1.1. EP Lubricant Wear Profile

Under EP lubrication, the wear values associated with the rollers, housing, and motorside mostly fall within the 1 to 200 range, while the wear cage displays range from 10 to approximately 500. This range has enhanced data dispersion that is considerably higher than other wear types within the EP category (Figure 8).

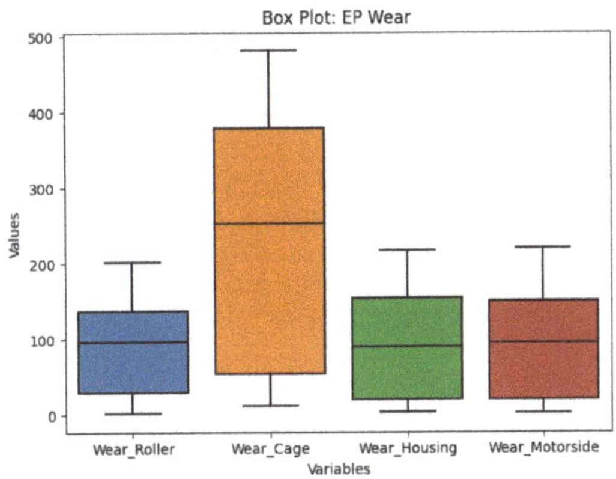

Figure 8. Box plot comparing the wear components of EP.

4.1.2. AW Lubricant Wear Profile

The wear characteristics of rollers, housing, and the motorside in AW lubricants typically fall between 0 and about 150. On the other hand, the wear cage varies from 30 to 350; however, this variation is not as pronounced as with EP lubricants. Notably, the wear roller in AW exhibits lower wear than the other types, as 50% of the data fall between 0 to approximately 70. The chemical codes AW6a and AW10a stand out as outliers since they fall beyond the average range indicated by the box plot's whiskers, which is another important point to make (Figure 9).

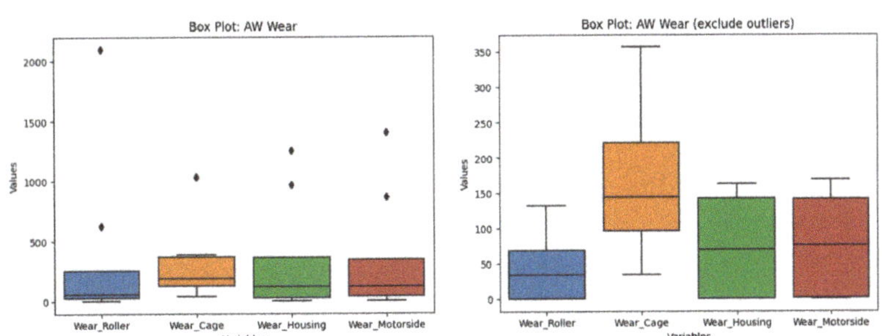

Figure 9. Box plot comparing the wear components of AW with outliers (dots in red rectangle shape) (**left**) and without outliers (**right**).

4.2. *Correlation between Average Wear and Chemical Predictors of EP and AW Lubricants*

We used correlation analysis in our study to evaluate the linear relationships between chemical predictors and average wear. The correlation coefficients (r) were calculated to provide an indication of the strength and direction of these interactions. An r value near 1 or −1 indicates a strong positive or negative connection, respectively, whereas values closer to 0 imply a weaker link. Positive correlation indicates that one variable increases as the other increases, and negative correlation means that, as one variable increases, the other decreases.

A significance level (α) of 0.05 was used to examine the statistical significance of the observed connections. This criterion indicates that there is a 5% chance we could reject the null hypothesis of no correlation when it is true (Type I error). In this analysis, correlations with p-values less than 0.05 were considered statistically significant, implying strong evidence against the null hypothesis.

To emphasis the most significant connections discovered in our data, we selected highlighted correlations that exceeded a magnitude of 0.6, representing a moderately strong correlation.

For activities ranging from 1 to 10, both CLPo_Add_L1 and Po_L1_S in AW indicate a positive and moderately strong correlation with average wear. Meanwhile, other chemical predictors display only weak correlations. However, they are not statistically significant (p-value < 0.05) (Figure 10). While the correlation coefficient indicates a relationship, the absence of statistical significance shows that conclusions should be drawn with caution. This means that while the correlation value suggests a relationship, the results are not strong enough (at the 5% significance level) to conclude that it is not simply caused by random variability in the data. AW's limited dataset size may have contributed to the high p-value.

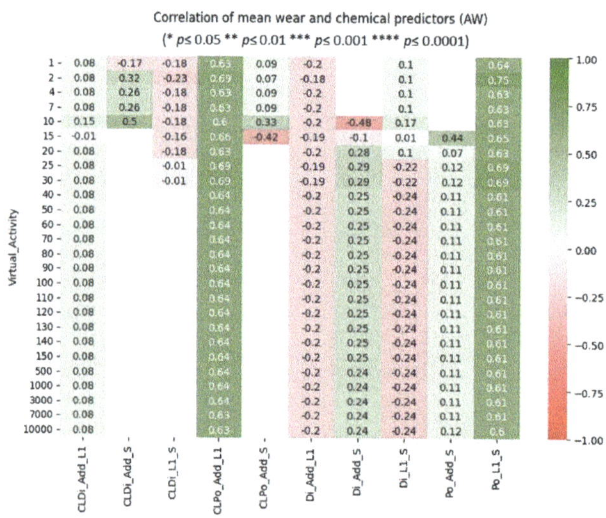

Figure 10. Correlation coefficient between average wear and chemical predictors of AW across different chemical activities (excluding outliers).

In the EP group, activity 4 is the only activity where Po_L1_S exhibits a positive and moderately strong relationship. Activity 2 shows that CLDi_L1_S and Di_Add_L1 have a negative, moderately strong correlation with average wear. A positive, moderately strong correlation is evident between average wear and both Di_Add_S (for activities 15–80) and Po_L1_S1 (for activities 4 and 15–10,000). Conversely, both Di_Add_L1 and Di_L1_S show a negative, moderately strong correlation between the average wear in activities 2, 4, and 15–10,000 and 15–10,000, respectively. Importantly, all these observed correlations are statistically significant (Figure 11).

A positive relationship with average wear suggests that as the values of these chemical predictors increase, the average wear also tends to increase. On the other hand, a negative relationship with average wear suggests that, as these chemical predictors increase, the average wear tends to decrease.

Furthermore, the intermediate strength of these correlations shows that these relationships are notable but not exactly linear.

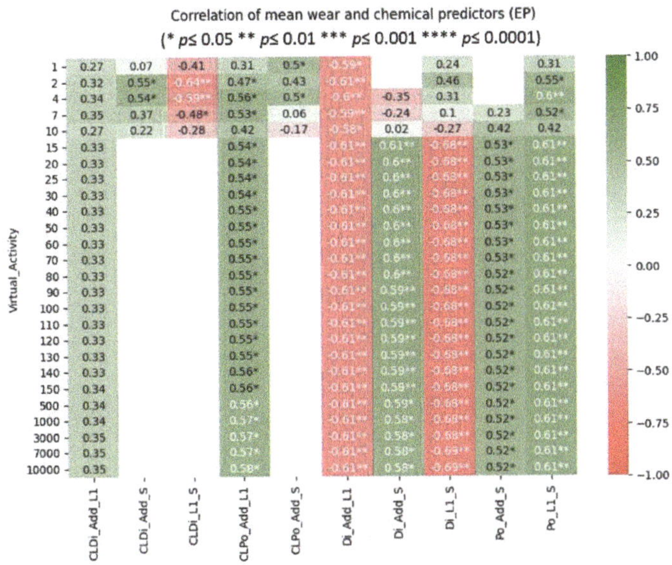

Figure 11. Correlation coefficient between the average wear and chemical predictors of EP across different chemical activities.

Finally, the fact that these correlations are statistically significant suggests that the observed relationships are not the result of random chance. The data show sufficient evidence against the null hypothesis.

4.3. Relationship of Average Wear and Mean of Cluster and Non-Cluster Chemical Predictors of EP and AW Lubricants:

In the EP category in Figure 12, we computed the average for the attributes of both cluster and non-cluster chemical predictors for each chemical code. When plotted against their average wear, the following distinct patterns emerged:

- The cluster and non-cluster groups scatter in contrasting directions, almost as if they mirror each other.
- There is a dramatic shift in patterns from activity 1 to activity 15. Beyond that, from activity 20 to 10,000, the changes are subtle.
- In activity 1, the non-cluster group stays on the left side of the graph. As activities progress, it moves gradually to the right. Conversely, the cluster group follows mirrored behaviour, initially starting on the right and moving leftward. The side switch can be observed from activity 7.
- Low-wear and high-wear groups can be observed in both cluster and non-cluster groups where data points above the 100 mean wear are considered high wear and below this is low wear.

In the AW category in Figure 13, we applied the same method as in EP and obtained the same pattern. However, there were some differences in AW.

- There is a dramatic shift in patterns in activities 1, 10 and 15.
- The side switch starts in activity 10, which is one activity after EP.

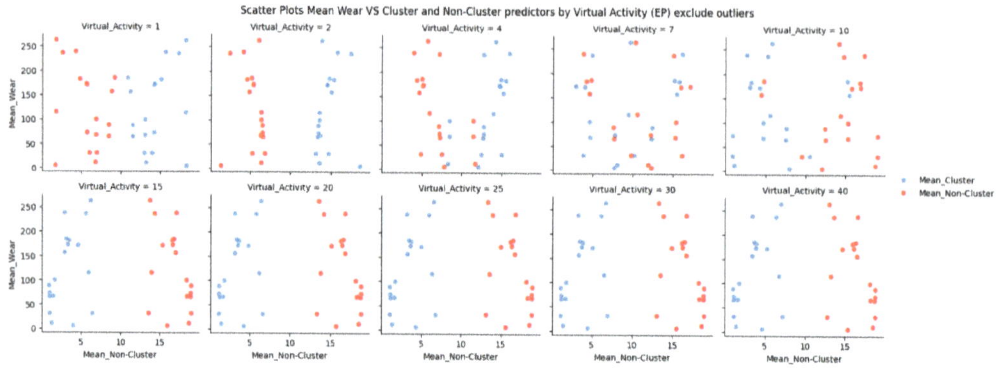

Figure 12. Scatter plot of average mean and average attributes of cluster and non-cluster chemical predictors across chemical activity 1–40.

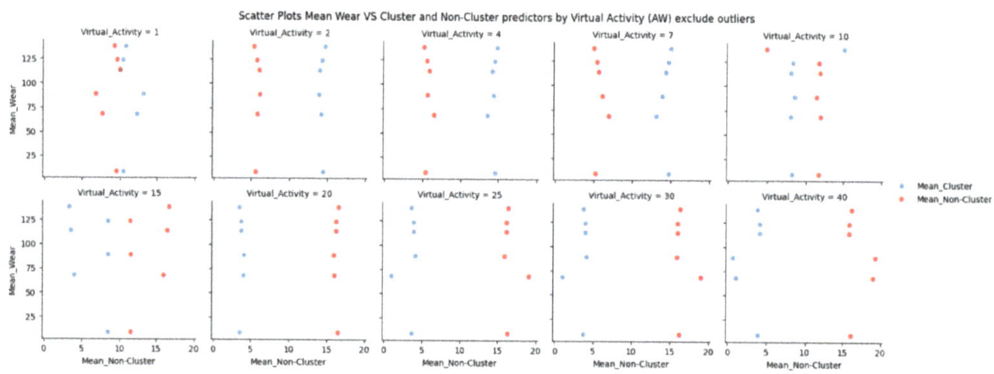

Figure 13. Scatter plot of average mean and average attributes of cluster and non-cluster chemical predictors in each chemical activity (AW—excluding outliers).

4.4. Relationship of Average Wear and Chemical Predictors of EP and AW Lubricants:

In Figure 14, Di_Add_L1 exclusively resides in the low-wear zone and exhibits a negative correlation with average wear. By contrast, other chemical predictors are present in both the high- and low-wear zones. Notably, 'CLDi_Add_S' is observed in activities 1, 2, 4, and 7, while both 'CLPo_Add_S' and 'CLDi_L1_S' appear in activities 1, 2, 4, 7, and 10.

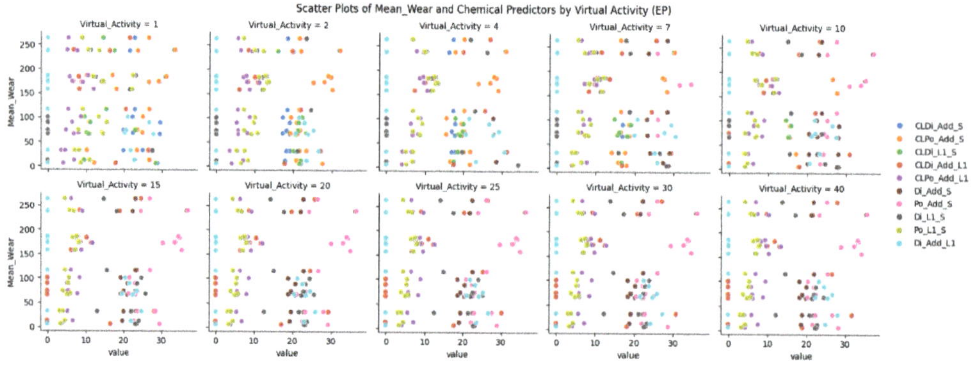

Figure 14. Scatter plot for average wear and chemical predictors in each chemical activity (EP).

For AW (Figure 15), Di_Add_L1 had a similar trend to EP, which exclusively resides in the low-wear zone with a negative relationship with the average wear except in activities 2,15,25, and 30, which show a positive relationship. Additionally, CLDi_Add_L1 only appears in the high-wear and AW group.

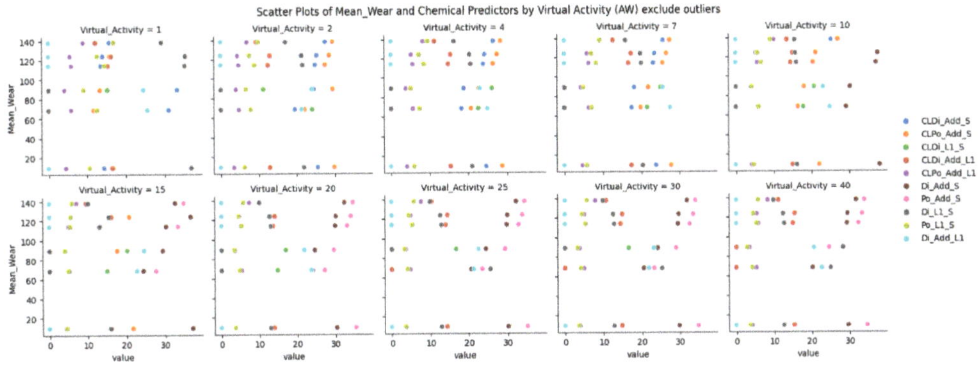

Figure 15. Scatter plot for average wear and chemical predictors in each chemical activity (AW) excluding outliers.

The differences in oil codes between the low- and high-wear groups are explained in Figure 16. The blue and green oil codes, which stand for Trimethylolepropaneoctoate (TMP) and PAO8 + TMP, respectively, are most frequently found in the high-wear category. The presence of the orange, purple, and red oil codes, which stand for PAO8, PAO40 + PAO8, and PAO40 + TMP, respectively, distinguished the low-wear group by contrast. Notably, CLDi_L1_S is primarily found in the low-wear zone, while Di_Add_L1 is only present there. Nevertheless, both wear zones are covered by chemical predictors, such as CLDi_Add_S, CLPo_Add_S, CLDi_Add_L1, CLPo_Add_L1, Di_Add_S, Po_Add_S, Di_L1_S, and Po_L1_S.

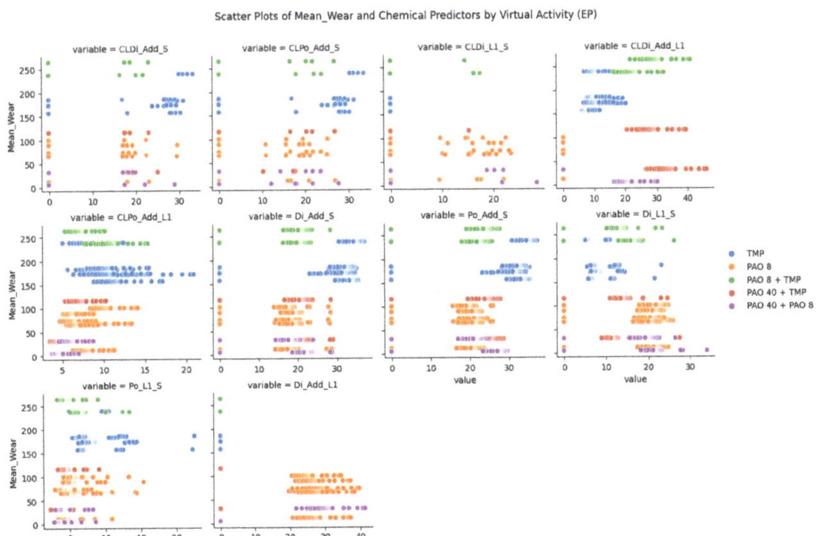

Figure 16. Scatter plot of average wear and chemical predictors grouped by oil code (EP).

Figure 17 shows that for AW, TMP and PAO 8 are the only two oil codes in AW. In particular, the connection between Di_Add_L1 and CLDi_L1_S is constant and is only observed in the low-wear category. This suggests that increasing the predictor has no effect on the average wear; instead, it stays the same.

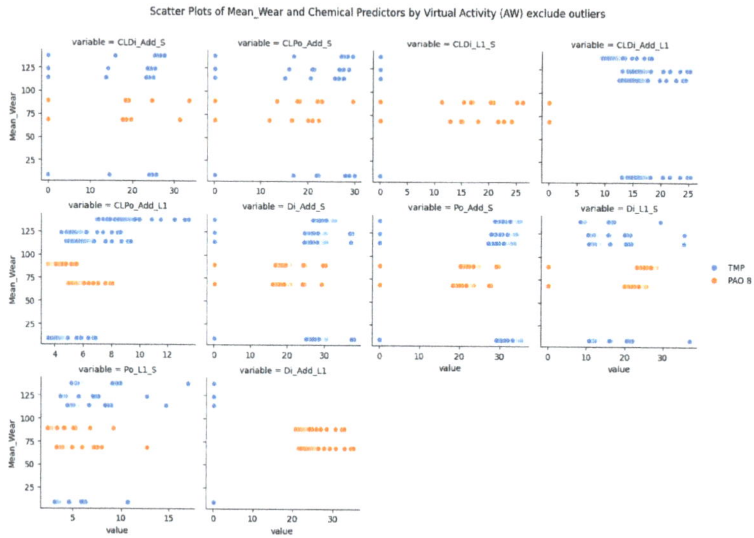

Figure 17. Scatter plot of average wear and each chemical predictor (AW) grouped by oil code (excluding outliers).

4.5. Relationship of Average Wear and Chemical Predictors of AW Lubricants with outliers:

On the other hand, if outliers are taken into account, Figure 18 shows that Di_Add_L1 and CLDi_L1_S appear in the high-wear range with the outliers, although they still show the same consistent association.

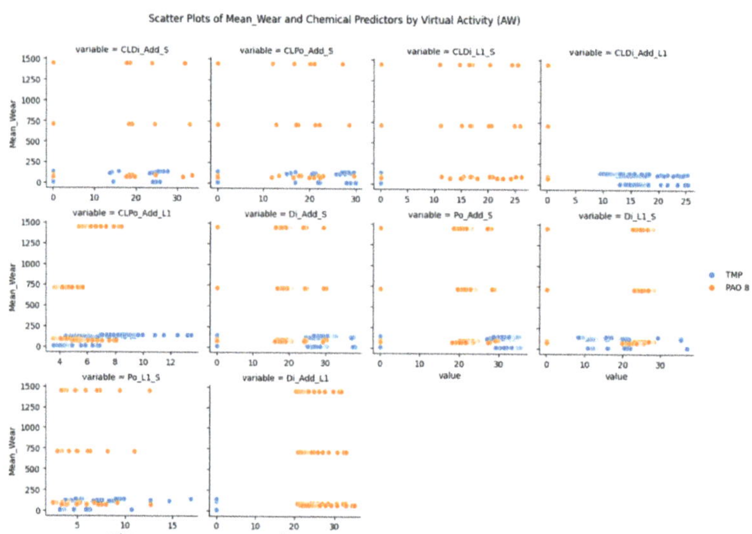

Figure 18. Scatter plot of average wear and each chemical predictor (AW) grouped by oil code (including outliers).

The correlation coefficient for the AW group, as shown in Figure 19, highlights a difference when outliers are taken into account: the correlations between average wear and chemical predictors for AW and EP shift in the reverse directions, but AW still shows no statistically significant result.

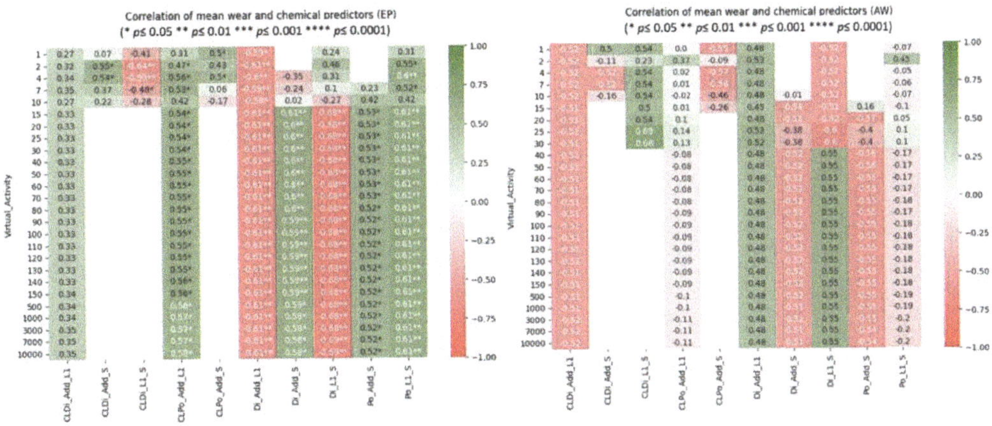

Figure 19. Correlation coefficient of average wear and chemical predictors of EP (**left**) and AW (**right**) across chemical activity (including outliers).

4.6. Logistics Regression to Observe Relationship of Wear Category and Chemical Predictors

As we classified the wear into low and high categories, we utilised g logistic regression to determine how chemical predictors influence these distinct wear levels. Notably, the logistic regression analysis is exclusively applied to the EP group, as AW did not yield statistically significant results.

To address issues with multicollinearity, we chose to include only three predictors (CLPo_Add_L1, Di_L1_S, and Di_Add_L1) in our model. These three predictors were chosen due to their low variance inflation factors (VIFs) (Figure 20), which indicates insignificant collinearity issues. In addition, their correlation coefficients with the wear category are the highest compared to those of the other variables (Figure 21).

	features	vif_Factor
0	Di_L1_S	3.517029
1	Di_Add_L1	1.848091
2	CLPo_Add_L1	2.510191

Figure 20. VIF factor.

This logistic regression model shows that "CLPo_Add_L1" and "Di_Add_L1" have statistically significant positive and negative influences, respectively, on the probability of being in the high wear category, while "Di_L1_S" has a statistically significant negative influence (Figure 22).

For each unit increase in "CLPo_Add_L1," log-odds of being in the high wear category increased by 1.9086 units ($p < 0.001$). An increase in one unit in "Di_L1_S" was associated with a decrease of approximately 0.0725 units in log-odds of the high wear category ($p < 0.05$). Conversely, for every unit increase in "Di_Add_L1," log-odds of being in the high wear category decreased by approximately 0.3294 units ($p < 0.001$).

	Di_Add_S	CLDi_Add_S	Po_Add_S	CLPo_Add_S	Di_L1_S	CLDi_L1_S	Po_L1_S	Di_Add_L1	CLDi_Add_L1	CLPo_Add_L1	Wear_Category
Di_Add_S	1.000000	-0.771182	0.864078	-0.698968	-0.212187	-0.386296	-0.315429	-0.332300	0.102619	0.242703	0.313114
CLDi_Add_S	-0.771182	1.000000	-0.754725	0.918503	-0.224561	0.423137	0.709835	-0.093044	-0.059851	-0.125212	0.064913
Po_Add_S	0.864078	-0.754725	1.000000	-0.847200	-0.074531	-0.516812	-0.408190	-0.296951	0.142914	0.215827	0.269441
CLPo_Add_S	-0.698968	0.918503	-0.847200	1.000000	-0.236496	0.497046	0.655329	-0.077717	-0.058570	-0.170602	0.030483
Di_L1_S	-0.212187	-0.224561	-0.074531	-0.236496	1.000000	-0.522634	-0.417116	0.419777	-0.176877	-0.290854	-0.441918
CLDi_L1_S	-0.386296	0.423137	-0.516812	0.497046	-0.522634	1.000000	0.320575	0.196730	-0.197674	-0.179072	-0.193248
Po_L1_S	-0.315429	0.709835	-0.408190	0.655329	-0.417116	0.320575	1.000000	-0.252430	-0.182260	0.275207	0.401795
Di_Add_L1	-0.332300	-0.093044	-0.296951	-0.077717	0.419777	0.196730	-0.252430	1.000000	-0.756998	-0.182621	-0.665935
CLDi_Add_L1	0.102619	-0.059851	0.142914	-0.058570	-0.176877	-0.197674	-0.182260	-0.756998	1.000000	-0.032403	0.340616
CLPo_Add_L1	0.242703	-0.125212	0.215827	-0.170602	-0.290854	-0.179072	0.275207	-0.182621	-0.032403	1.000000	0.564637
Wear_Category	0.313114	0.064913	0.269441	0.030483	-0.441918	-0.193248	0.401795	-0.665935	0.340616	0.564637	1.000000

Figure 21. Correlation matrix (red for negative and green for positive, the intensity represents the magnitude).

```
Optimization terminated successfully.
         Current function value: 0.140852
         Iterations 10
                           Results: Logit
=================================================================
Model:              Logit            Method:           MLE
Dependent Variable: Wear_Category    Pseudo R-squared: 0.795
Date:               2023-08-24 03:10 AIC:              139.8372
No. Observations:   468              BIC:              156.4311
Df Model:           3                Log-Likelihood:   -65.919
Df Residuals:       464              LL-Null:          -321.50
Converged:          1.0000           LLR p-value:      1.8211e-110
No. Iterations:     10.0000          Scale:            1.0000
-----------------------------------------------------------------
               Coef.  Std.Err.    z     P>|z|    [0.025   0.975]
-----------------------------------------------------------------
const         -9.7268   1.4384 -6.7624 0.0000 -12.5460 -6.9077
CLPo_Add_L1    1.9086   0.2549  7.4867 0.0000   1.4090  2.4083
Di_L1_S       -0.0725   0.0286 -2.5307 0.0114  -0.1286 -0.0163
Di_Add_L1     -0.3294   0.0448 -7.3544 0.0000  -0.4171 -0.2416
=================================================================
```

Figure 22. Logistic regression result (EP).

These findings illuminate the relationships between these chemical predictors and wear patterns. The statistically significant coefficients indicate that "CLPo_Add_L1" positively influences high wear, whereas "Di_L1_S" and "Di_Add_L1" negatively influence high wear in EP lubricants. "CLPo_Add_L1" had a statistically significant positive coefficient, showing that higher quantities of this chemical predictor are related to an increased chance of high wear in EP lubricants. By contrast, the predictors "Di_L1_S" and "Di_Add_L1" had statistically significant negative coefficients, indicating that higher quantities of these chemical predictors are associated with a reduced likelihood of high wear.

In conclusion, our logistic regression analysis reveals the significant influence of these chemical predictors on wear categories, casting light on the factors that contribute to wear patterns in our study.

4.7. Other Significant Findings

The AW group, without outliers, only shows a significant correlation in the wear roller. CLPo_Add_L1 shows a strong positive relationship with the wear roller for all activities. By contrast, 'Po_L1_S' demonstrates this strength of association only in activities 1–15, whereas 'CLDi_Add_S' demonstrates a similar strength of association only in activity 10 (Figure 23).

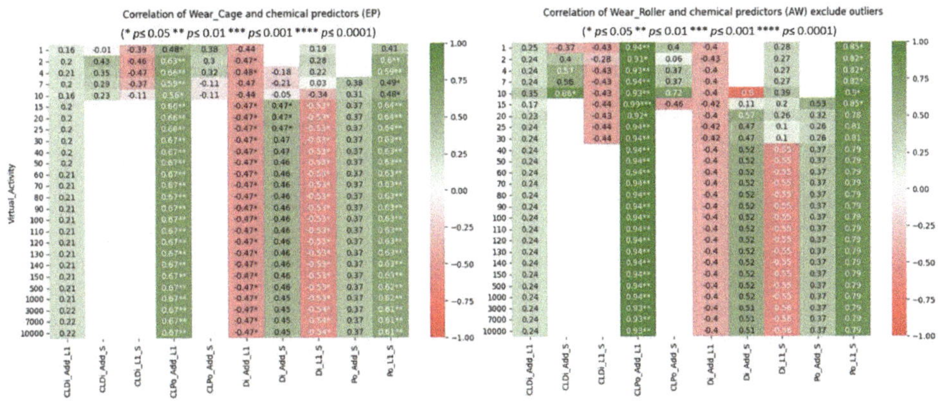

Figure 23. Correlation of wear cage and EP predictors (**left**) and wear roller and AW predictors excluding outliers (**right**).

Given the high correlation observed with CLPo_Add_L1, implying a strong linear relationship, we proceeded to conduct a linear regression analysis with this predictor (see Figure 24). Additionally, we included Po_L1_S as it contributed to an improved R-squared value. The results of this regression analysis yielded an adjusted R-squared of 0.67. This figure indicates that the selected independent variables in the model, such as "CLPo_Add_L1" and "Po_L1_S," explain approximately 66% of the variation in the dependent variable, "Wear_Roller." Notably, some prospective predictors were excluded from the model on purpose. Their lack of statistically significant relationships with the dependent variable and their inability to enhance the adjusted R-squared value led to this conclusion.

```
                            OLS Regression Results
==============================================================================
Dep. Variable:            Wear_Roller   R-squared:                       0.667
Model:                            OLS   Adj. R-squared:                  0.663
Method:                 Least Squares   F-statistic:                     153.2
Date:                Thu, 24 Aug 2023   Prob (F-statistic):           2.95e-37
Time:                        04:30:04   Log-Likelihood:                -730.50
No. Observations:                 156   AIC:                             1467.
Df Residuals:                     153   BIC:                             1476.
Df Model:                           2
Covariance Type:            nonrobust
===============================================================================
                  coef    std err          t      P>|t|      [0.025      0.975]
-------------------------------------------------------------------------------
const         -80.5257      7.624    -10.562      0.000     -95.587     -65.464
CLPo_Add_L1    17.0164      1.096     15.533      0.000      14.852      19.181
Po_L1_S         4.3316      0.980      4.419      0.000       2.395       6.268
==============================================================================
Omnibus:                        1.058   Durbin-Watson:                   0.273
Prob(Omnibus):                  0.589   Jarque-Bera (JB):                1.108
Skew:                           0.192   Prob(JB):                        0.575
Kurtosis:                       2.850   Cond. No.                         29.9
==============================================================================
```

Figure 24. Linear regression wear roller—AW excluding outliers.

Interpreting the coefficients, for every one unit of change in CLPo_Add_L1, there was an estimated change of 17.0164 units in the wear roller. Similarly, a one-unit change in Po_L1_S corresponded to an estimated change of 4.33 units in the wear roller. These findings illuminate the impact of both CLPo_Add_L1 and Po_L1_S on the wear roller in the AW group.

5. Conclusions

Commonly accepted hypotheses in tribology state that, in the case of boundary lubrication, functional additives prevent direct metal–metal contact in the contact zone of the machinery element. Traditionally, boundary lubrication is associated with the relation of the film height in the contact toward the height of the asperities. The film height is calculated traditionally using the theory of elastohydrodynamics (EHL), stating that the viscosity (e.g., film thickness) locally present in the contact zone is the prime predictor defining a boundary situation. Classical EHL states that viscosity, as the relevant predictor, transforms using the local energy impact in the contact zone given by temperature, temperature flashes, and shear impact, both leading to a decay in the viscosity. By contrast, the pressure assumed to be locally present in the contact zone compresses the lubricant, which is associated with an increase in viscosity in the lubricant by means of a viscosity-pressure coefficient. A lot of computational effort is carried out in order to obtain an insight into these processes and the structure–property relationship given by the chemical structure of the lubricant and their ingredients. However, up until now, no specific rules have been seen regarding how wear processes are assumed to prevent, but in some cases, are also found to promote and are predictable. The transit of EHL toward thin film lubrication (TFL) and tribocatalysis (TC) is accompanied by various physical transitions within the lubricant, notified as their "inner" structure herein. The exposure to the impact of tribological stress causes the lubricant to transform its spatial solute structure before it starts to react chemically. These structural changes are described within this paper by calculating the chemical potential, differentiated in the dipolar and induced dipolar (polarisability) activity appearing at the surface (S) and in the solvent (L). Solutes in solvents and at surfaces may create clusters under stress in order to dissipate it. This study permuted all different structural rearrangements of the lubricants in relation to imposed external activity. This activity could arise from temperature, pressure, and surfaces becoming active due to the tribological impact. The study here applies the assumption comprising the formation of "inner" structures, clusters (CL), and non-clusters to a variety of lubricants with well-known chemical structures and their ability to prevent or induce wear using the bearing test rig FE8 (DIN 51819). The lubricants were divided into two different base oils (Poly-a-Olefine (PAO) and an Ester (Trimethylpropane Octoate (TMP) and were placed under different extreme pressure (EP) and anti-wear (AW) additives. The wear within this test is scattered over the single constituents, e.g., cage, rollers, and washer. For all the different lubricants, the "inner" structure was calculated as dipolar and induced dipolar activity and their likelihood to form clusters at the surface and within the base oil. A detailed statistical investigation, searching for relations of the "inner" structure with individuals, but also the average wear led to the following results:

The base oil, rather than the additives, significantly guides the appearance of high or low wear.

The lubricant predictors change abruptly due to the increase in the surface activity. This leads to the conclusion that determining the activity of a lubricant during any test requires a careful definition of how the load capacity of the component is related to the load capacity of the lubricant. This is to say, the result of any load-related tribological test may abruptly change under different test conditions.

The encapsulation of induced dipole moments (clustered polarisability) and additives in a solvent are relevant as clusters promote significant wear. We are interpreting this result by the fact that a wear process is highly transient by exchanging electrons, charged and uncharged particles. These transient exchange processes are scavenged if the lubricant uptakes them via charge dissipation due to its polarisation ability. If the "inner" structure of the lubricant clusters the polarisability within the solvent, this functionality is blocked, and the wear process is not dissipated in the lubricant.

The presence of outliers in the AW dataset, combined with its small size, suggests the unfavourable effect on statistical metrics in smaller datasets, thus misleading the interpretation of the results. The situation emphasises the importance of dataset size and quality in tribological research. It raises the delicate problem of how to approach outliers, particularly in smaller datasets, where their influence can be clearer. The topic for discussion is whether outliers should be thoroughly accounted for in order to maintain the dataset's accuracy or whether they should be set aside because they may represent unique but real-world conditions. This opens the door to a broader discussion about improving data-gathering procedures in tribology. Future research could focus on increasing the AW dataset to either confirm or refine the current study's conclusions while carefully examining the influence of outliers in the analysis.

6. Summary

The transition from EHL toward TFL and TC requires a detailed understanding of how lubricants uptake the exposition to contact stress. Apart from viscosity-related parameters, the "inner" structure of a lubricant comes into the picture. Despite the huge variety of chemicals, the simple calculation of the dipolar and induced dipolar structure leads to powerful (e.g., robust) predictors for wear under thin film lubrication (TFL) conditions. Despite the fact that the chemical structures had to be assumed due to their technical grade, the prediction is seemingly very robust if the extreme pressure additives are taken. This insight is even more important as new lubricants with respect to the prediction of wear have to be found for E-mobility and ecology purposes.

Author Contributions: Conceptualization, W.H. and L.K.; methodology, W.H. and L.K.; software, W.H. and L.K.; validation, W.H. and B.G.; formal analysis, L.K.; investigation, W.H.; resources, W.H.; data curation, L.K.; writing—original draft preparation, L.K.; writing—review and editing, W.H. and L.K.; visualization, W.H. and L.K.; supervision, J.F.; project administration, J.F.; funding acquisition, B.G. All authors have read and agreed to the published version of the manuscript.

Funding: This research received no external funding.

Data Availability Statement: The data presented in this study are available on request from the corresponding author. The data are not publicly available due to privacy restrictions.

Conflicts of Interest: The authors declare no conflict of interest.

Appendix A

List of Additives

EP

1. Triphenylphosphorothionate
2. Tridodecyltrisulfide
3. Sulfurated Isobutene S2Cl2 Route
4. Sulfurated Isobutene NaSx Route
5. Sulfurated Rapseed Oil
6. Dithiophosphoric Acid Isopropylester

AW

1. Methylenebisdithiocarbamate
2. Molybdenumbisdithiocarbamate
3. Zinc-2-ethylhexyldithiophosphate
4. Molybdenum-2-ethylhexyldithiophosphate
5. Tricresylphosphate
6. 2-Ethylhexyldithiophosphate-Amine neutralised
7. Bis-C13-alkylammonium -dihexylphosphate
8. 2.6-Dimercaptodithiazole

References

1. Bushan, B. *Introduction to Tribology*; Wiley: Hoboken, NJ, USA, 2013; ISBN 978-1-119-94453-9.
2. Moore, D.F. *Principals and Applications of Tribology*; Pergamon Press: Elmsford, UK, 1975; ISBN 0-08-0179-020.
3. Shizhu, W.; Ping, H. *Principles of Tribology*; John Wiley & Sons: Hoboken, NJ, USA; Tsinghua University Press: Beijing, China, 2012.
4. Deters, L. Reibung, Verschleiß und Schmierung. In *Konstruktionselemente des Maschinenbaus 2*; Sauer, B., Ed.; Springer: Berlin/Heidelberg, Germany, 2008; pp. 1–68. ISBN 978-3-540-76653-7.
5. Shi, J.; Zhao, B.; He, T.; Tu, L.; Lu, X.; Xu, H. Tribology and dynamic characteristics of textured journal-thrust coupled bearing considering thermal and pressure coupled effects. *Tribol. Int.* **2023**, *180*, 108292. [CrossRef]

6. Bowden, F.P. The Friction And Lubrication of Solids. In *Oxford Classsic Texts in the Physical Sciences;* Oxford University Press: Oxford, UK, 2001; ISBN 13-978-0198507772.
7. Priest, M.; Ehret, P.; Flamand, L.; Dalmaz, G.; Childs TH, C.; Dowson, D.; Berthier, Y.; Taylor, C.M.; Lubrecht, A.A. *Lubrication at the Frontier: The Role of the Interface and Surface Layers in the Thin Film and Boundary Regime*, 1st ed.; Elsevier Science: Amsterdam, The Netherlands, 1999; ISBN 9780080535661.
8. Zhu, D.; Wang, Q.J. Elastohydrodynamic Lubrication: A Gateway to Interfacial Mechanics—Review and Prospect. *J. Tribol.* **2011**, *133*, 041001. [CrossRef]
9. Wills, G. *Lubrication Fundamentals (Mechanical Engineering);* Marcel Dekker: New York, NY, USA, 1980.
10. Ginsberg, J.H. *Advanced Engineering Dynamics;* Cambridge University Press: Cambridge, UK, 1995; ISBN 13 978-0521470216.
11. Zhang, Y.; Biboulet, N.; Venner, C.H.; Lubrecht, A. Prediction of the Stribeck curve under full-film Elastohydrodynamic Lubrication. *Tribol. Int.* **2020**, *149*, 105569. [CrossRef]
12. Bi, Z.; Mueller, D.W.; Zhang, W. State of the art of friction modelling at interfaces subjected to elastohydrodynamic lubrication (EHL). *Friction* **2020**, *9*, 207–227. [CrossRef]
13. Ewen, J.P.; Spikes, H.A.; Dini, D. Contributions of Molecular Dynamics Simulations to elastohydrodynamic Lubrication. *Tribol. Lett.* **2021**, *69*, 24. [CrossRef]
14. Hamrock, B.J.; Steven, R.; Schmid, S.R.; Jacobson, B.O. *Fundamentals of Fluid Film Lubrication*, 2nd ed.; CRC Press Inc.: Boca Raton, FL, USA, 2004; ISBN 13 978-0824753719/10 0824753712.
15. Szeri, A.Z. *Fluid Film Lubrication: Theory and Design*, 2nd ed.; Cambridge University Press: Cambridge, UK, 2011; ISBN 13 978-0521898232/10 0521898234.
16. Zakharov, S. Hydrodynamic lubrication research: Current situation and future prospects. *J. Frict. Wear* **2010**, *31*, 56–67. [CrossRef]
17. Tan, Y.; Huang, W.; Wang, X. Tribochemistry of ZDDP in molecular orbital calculations. *Tribol. Int.* **2004**, *37*, 447–450. [CrossRef]
18. Khonsari, M.M.; Booser, R.E. *Applied Tribology: Bearing Design and Lubrication;* John Wiley & Sons, Ltd.: Hoboken, NJ, USA, 2008; ISBN 978-0-470-05711-7.
19. Kudish, I.I. *Elastohydrodynamic Lubrication for Line and Point Contacts;* CRC Press: Boca Raton, FL, USA, 2013; ISBN 978-1-4665-8390-0.
20. Spikes, H.A. Triboelectrochemistry: Influence of applied electrical potentials on friction and wear of lubricated contacts. *Tribol. Lett.* **2020**, *68*, 90. [CrossRef]
21. Li, X.; Tong, W.; Shi, J.; Chen, Y.; Zhang, Y.; An, Q. Tribocatalysis mechanisms: Electron transfer and transition. *J. Mater. Chem. A* **2023**, *11*, 4458–4472. [CrossRef]
22. Holweger, W. Novel Predictors for Friction and Wear in Drivetrain Applications. In *Friction, Lubrication and Wear;* Chowdhury, M.A., Ed.; Intechopen: London, UK, 2019.
23. Sommer, K.; Heinz, R.; Schöfer, J. *Verschleiß Metallischer Werkstoffe;* Springer Vieweg: Wiebaden, Germany, 2014; ISBN 978-3-8348-2464-6.
24. Hostis, B.L.; Minfray, C.; Frégonèse, M.; Verdu, C.; Ter-Ovanessian, B.; Vacher, B.; Le Mogne, T.; Jarnias, F.; D'Ambros, A.D.-C. Influence of lubricant formulation on rolling contact fatigue of gears—Interaction of lubricant additives with fatigue cracks. *Wear* **2017**, *382–383*, 113–122. [CrossRef]
25. Kovalchenko, M.V.; Korchak, S.N.; Kuznetsov, O.L.; Erdemir, A.G.V.; Samsonov, G.V. The effects of mechanical stresses on catalytic processes. *J. Catal.* **2002**, *212*, 232–235.
26. Erdemir, A. Tribocatalysis: A new field at the frontier of tribology and catalysis. *Surf. Sci.* **2014**, *631*, 2–8.
27. Chen, Z.; Zhang, R.; Gao, Y.; Yang, Q.; Liu, J. The role of catalysts in friction and wear. *Friction* **2019**, *7*, 271–293.
28. Dickinson, T.; Jensen, L.C.; Lee, S.; Scudiero, L.; Langford, S.C. Fracto-emission and electrical transients due to interfacial failure. *J. Adhes. Sci. Technol.* **1994**, *8*, 1285–1309. [CrossRef]
29. Dickinson, J.T.; Jensen, L.C.; Jahan-Latibari, A. Fracto-emission: The role of charge separation. *J. Vac. Sci. Technol. A* **1984**, *2*, 1112–1116. [CrossRef]
30. Kajdas, C.K. Importance of the triboemission process for tribochemical reaction. *Tribol. Int.* **2005**, *38*, 337–353. [CrossRef]
31. Huang, W.; Tan, Y.; Chen, B.; Dong, J.; Wang, X. The binding of antiwear additives to iron surfaces: Quantum chemical calculations and tribological tests. *Tribol. Int.* **2003**, *36*, 163–168. [CrossRef]
32. Jayadas, N.H.; Prabhakaran Nair, P. Study of the Anti-Wear Properties of Coconut Oil Using Quantum Chemical Calculations and Tribological Tests. *ASME J. Tribol.* **2006**, *128*, 654–659. [CrossRef]
33. Vakis, A.; Yastrebov, V.; Scheibert, J.; Nicola, L.; Dini, D.; Minfray, C.; Almqvist, A.; Paggi, M.; Lee, S.; Limbert, G.; et al. Modeling and simulation in tribology across scales: An overview. *Tribol. Int.* **2018**, *125*, 169–199. [CrossRef]
34. Evans, M.-H. An updated review: White etching cracks (WECs) and axial cracks in wind turbine gearbox bearings, Materials Science and Technology. *Mater. Sci. Technol.* **2016**, *32*, 1133–1169. [CrossRef]
35. Šmeļova, V.; Schwedt, A.; Wang, L.; Holweger, W.; Mayer, J. Microstructural changes in White Etching Cracks (WECs) and their relationship with those in Dark Etching Region (DER) and White Etching Bands (WEBs) due to Rolling Contact Fatigue (RCF). *Int. J. Fatigue* **2017**, *100*, 148–158. [CrossRef]
36. Spille, J.; Wranik, J.; Barteldes, S.; Mayer, J.; Schwedt, A.; Zürcher, M.; Lutz, T.; Wang, L.; Holweger, W. A study on the initiation processes of white etching cracks (WECs) in AISI 52100 bearing steel. *Wear* **2021**, *477*, 203864. [CrossRef]
37. Holweger, W.; Schwedt, A.; Rumpf, V.; Mayer, J.; Bohnert, C.; Wranik, J.; Spille, J.; Wang, L. A Study on Early Stages of White Etching Crack Formation under Full Lubrication Conditions. *Lubricants* **2022**, *10*, 24. [CrossRef]

38. Esmaeili, K.; Wang, L.; Harvey, T.J.; White, N.M.; Holweger, W. A Study on the Influence of Electrical Discharges on the Formation of White Etching Cracks in Oil-Lubricated Rolling Contacts and Their Detection Using Electrostatic Sensing Technique. *Lubricants* **2023**, *11*, 164. [CrossRef]
39. Wranik, J.; Holweger, W.; Lutz, T.; Albrecht, P.; Reichel, B.; Wang, L. A Study on Decisive Early Stages in White Etching Crack Formation Induced by Lubrication. *Lubricants* **2022**, *10*, 96. [CrossRef]
40. Gould, B.; Demas, N.; Erck, R.; Lorenzo-Martin, M.C.; Ajayi, O.; Greco, A. The Effect of Electrical Current on Premature Failures and Microstructural Degradation in Bearing Steel. *Int. J. Fatigue* **2021**, *145*, 106078. [CrossRef]
41. Shivakumar, D.; Williams, J.; Wu, Y.; Damm, W.; Shelley, J.; Sherman, W. Prediction of Absolute Solvation Free Energies using Molecular Dynamics Free Energy Perturbation and the OPLS Force Field. *J. Chem. Theory Comput.* **2010**, *6*, 1509–1519. [CrossRef]
42. Mezei, M. The finite difference thermodynamic integration, tested on calculating the hydration free energy difference between acetone and dimethylamine in water. *J. Chem. Phys.* **1987**, *86*, 7084–7088. [CrossRef]
43. Holweger, W.; Bobbio, L.; Mo, Z.; Fliege, J.; Goerlach, B.; Simon, B. A Computational Study on the Role of Lubricants under Boundary Lubrication. *Lubricants* **2023**, *11*, 80. [CrossRef]
44. Albright, S.C.; Winston, W.L. Business Analytics: Data analysis & decision making. In *Describing the Distribution of a Single Variable*; Cengage Learning, Inc.: Beijing, China, 2016; pp. 49–51.

lubricants MDPI

Article

A Methodological Approach to Assessing the Tribological Properties of Lubricants Using a Four-Ball Tribometer

Tareq M. A. Al-Quraan [1,*], Fadi Alfaqs [2], Jamil Haddad [2], Viktor Vojtov [3], Anton Voitov [3], Andrey Kravtsov [3], Oleksandr Miroshnyk [4,*] and Andrii Kondratiev [5]

[1] Department of Technical Sciences, Ma'an College, Al-Balqa Applied University, Al-Salt 19117, Jordan
[2] Department Mechanical Engineering, Faculty of Engineering Technology, Al-Balqa Applied University, Amman 11134, Jordan; faalfaqs@bau.edu.jo (F.A.); drjamil@bau.edu.jo (J.H.)
[3] Department of Transport Technologies and Logistics, State Biotechnological University, 61052 Kharkiv, Ukraine; vavoitovva@btu.kharkiv.ua (V.V.); klkavoitov@btu.kharkiv.ua (A.V.); kravcov_84@btu.kharkiv.ua (A.K.)
[4] Department of Electricity Supply and Energy Management, State Biotechnological University, 61052 Kharkiv, Ukraine
[5] Department of Materials Science and Engineering of Composite Structures, O.M. Beketov National University of Urban Economy in Kharkiv, Marshal Bazhanov Str. 17, 61002 Kharkiv, Ukraine; andrii.kondratiev@kname.edu.ua
* Correspondence: tarjorggg@bau.edu.jo (T.M.A.A.-Q.); omiroshnyk@btu.kharkiv.ua (O.M.)

Abstract: Based on the analysis of standards for the testing of lubricants, both liquid and plastic, on a four-ball tribometer, and the analysis of the parameters by which lubricants are evaluated, this paper proposes a methodology and an integral parameter for the estimation of tribological properties. The methodological approach proposed in this paper allows for the integration of a variety of parameters provided in the standards for the testing of lubricants into one indicator. Herein, we show that the developed technique is based on the energy approach and takes into account the specific wear work of the test material (steel balls) in the lubricating medium to be investigated. The results of laboratory tests of a wide range of lubricants are presented: hydraulic fluids, motor and transmission oils of various purposes and classifications. It is shown that the magnitude of the integral parameter can be used to assess the effectiveness of anti-wear and anti-scuff additives in base lubricants, as well as the ranges of their applications. This allows for differentiation and quantitative evaluation of the effectiveness of such additives. The obtained results allow us to state that all tests according to the developed method are reproducible and homogeneous, which is confirmed using the Cochran criterion. The coefficient of variation during testing does not exceed 18%. We show that the presented methodology and the integral parameter can be used in the first stage of the laboratory selection tests of new lubricants and additives of various origins, reducing the costs of their development and implementation.

Keywords: lubricants; hydraulic oils; motor oils; gear oils; four-ball tribometer; anti-wear additives; extreme pressure additives

1. Introduction

An analysis of publications published in various scientific journals allows us to conclude that, recently, new types of lubricant additives and dopants have been actively developed. Much attention has been paid to additives based on nanomaterials, as well as environmentally friendly lubricants based on vegetable oils. The search for the optimal concentrations of additives, the use of additives in the base lubricant and the evaluation of the effectiveness of such additives require a large number of laboratory, standard and operational tests.

In order to reduce the time and costs of such studies, developers use laboratory tests on a four-ball tribometer. Such tests are the first stage of research and allow for the rapid evaluation of the effectiveness of lubricants, the establishment of optimal concentrations and justification for areas of research.

The second stage is the testing of new lubricants on real tribo-system constructions in the form of standard tests or controlled operation. Therefore, the use of a four-ball tribometer in the first stage of research can reduce the time required for testing and increase the efficiency of such studies.

The methodology for the study of lubricants using a four-ball tribometer is set out in the standards of various countries: GOST 9490, ASTM D4172, ASTM D2596, ASTM D2783, ASTM D5183 and DIN 51350. The standards allow for the determination of several parameters, the physical meanings of which differ and are sometimes contradictory, making it difficult to formulate specific conclusions. In our opinion, the tribological properties of a lubricant should be evaluated by an integral indicator that takes into account the interaction of the lubricating medium (additives and dopants in the base lubricating medium) with friction surface materials under various loads, revealing the mechanisms of physical adsorption and chemisorption and the chemical reactions that affect the wear resistance of the tribo-system. Such an approach can increase the efficiency of the development of new lubricants and make the test procedure on a four-ball tribometer more informative.

An overview of various designs of tribometers that are used in the laboratory testing of materials and lubricants for friction and wear is presented in [1]. Among the variety of test equipment constructions, specific consideration is given to the four-ball tribometer. This device is used to determine the tribological characteristics of lubricants, both liquid and plastic.

According to the authors of [2], traditional tests on a four-ball tribometer make it possible to optimize base lubricant additives and dopants through an assessment of their effective actions. We believe that such a methodological approach makes it possible to reduce the number of preliminary tests and to detail the work of extreme pressure additives.

The authors of [3] used a four-ball tribometer to test the extreme pressure properties of lubricants according to ASTM D2596. The authors investigated various lubricants and their ability to prevent burrs. A feature of the tests was the change in the rotational speed of the tribometer rotor. The authors found that the change in the rotor speed and the acceleration time of the rotor changed the magnitude of the burr load. In our opinion, these studies show the necessity of meeting the requirements of the standards for testing using a four-ball tribometer, which set out the testing procedure. Otherwise, the results obtained in different laboratories and by different researchers will vary significantly, making it impossible to analyze such data and create databases. Changes to the methods set out in the standards should be substantiated.

The anti-wear and extreme pressure properties of motor oils were studied in [4–6] using a four-ball tribometer according to ASTM D4172 and ASTM D2783. The tribological characteristics of fresh and used SAE 15W40 and SAE 20W50 engine oils were evaluated at different operating times in the engine [5]. The presented results allow us to state that the chosen test method is highly informative. This method allows for the evaluation of the effectiveness of additives in motor oils, such as colloidal particles and ionic liquids [4], as well as carbon nanotubes [6]. These works confirm the high sensitivity and information contents of tests conducted on a four-ball tribometer.

A study of the efficiency of the use of graphene-based organic additives based on graphite, which provide a tribo-chemical reaction during friction with metal surfaces, was described in [7]. The authors of the work showed the high efficiency of using a four-ball tribometer in the study of such additives. According to the test results, organic additives are physically or chemically adsorbed on rubbing metal surfaces with the formation of monolayers, which indicates the high information content of the chosen method.

The results of tests of greases conducted on a four-ball tribometer were presented in [8–10]. For example, the addition of ash from rice husk was shown in [8]. In [9,10], additives to lithium grease in the form of talc nanoparticles were studied. The main conclusion of the works listed above was that tests using a four-ball tribometer can achieve highly reproducible results.

Therefore, we can state that the use of a four-ball tribometer in the first stage of the qualifying tests of oils, lubricants, additives and dopants can help to obtain reliable information about the selected areas of research, introducing corrections into research in the first stages of testing and reducing the financial costs and time spent on controlled exploitation [9–12].

A good result was achieved using a four-ball tribometer in the study of the tribological characteristics of vegetable oils. The investigation of sunflower vegetable oil with nano-additives at various concentrations was shown in [13], which allowed the authors to establish the optimal concentration. The purpose of the studies presented in the article was to assess the anti-wear characteristics of vegetable-based lubricants. The test results from the presented work enable us to assert the high information value of employing a four-ball tribometer in studying the anti-wear and extreme pressure properties of vegetable-based lubricants.

The evaluation of the tribological characteristics of various lubricants, both liquid and plastic, with the addition of nanoparticles, where the main tool is a four-ball tribometer, was described in [14–21]. The tribological properties of SiO_2 nanoparticles used as additives to a base oil were studied in [15]. In this work, the authors used a four-ball tribometer to evaluate the anti-wear properties of the base oil in comparison with SiO_2 nanoparticles in the base oil. A strong correlation was observed between the anti-wear properties of oils with additives, their solubility, and the additive content in the base oil. This affirms the high information value of the chosen research method [16].

The tribological properties of a paraffin lubricant with ZnO and CuO nanoparticles were studied in [17]. Using a four-ball tribometer, the extreme pressure and anti-wear properties of lubricants with nano-additives were assessed according to the ASTM D2596 and ASTM D2266 standards.

In [18], the tribological performance of castor grease with and without two-dimensional (2D) lamellar nanomaterials was evaluated using a four-ball tribo-tester following the ASTM standards. The test results demonstrated high measurement reproducibility.

Nanoparticles, specifically activated carbon, in lithium grease, were discussed in [19,20], while hybrid nano-oils were studied in [21,22]. The possibility of using nanoparticles and polytetrafluoroethylene simultaneously was considered by the authors of [23].

The tribological characteristics of lubricants containing carbon spheres were discussed in [24]. In all these works, experimental studies were conducted using a four-ball tribometer. The authors of these works highlight the high reproducibility of the experimental results.

The works in [25,26] consisted of experimental studies on the tribological characteristics of lubricants using a four-ball tribometer with polytetrafluoroethylene additives. Appropriate concentrations of polytetrafluoroethylene additives in base oils were established.

All the works mentioned relied on tests conducted using a four-ball tribometer [14–26]. In [27,28], the authors proposed an energy parameter, namely the specific work of the wear of a test material composed of ball-bearing steel in the tested lubricating medium, to assess the tribological characteristics of lubricants. The study demonstrated that the value of the specific wear work can serve as a comprehensive energy parameter for the evaluation of the lubricating properties of liquid and plastic materials. The presented methodology and parameters for the assessment of the effectiveness of anti-wear and extreme pressure additives include the mandatory determination of the wear index and its inclusion in the general formula. In our opinion, this methodological approach reduces the information content of the test results obtained from a four-ball tribometer [24,29,30]. Excluding the wear indicator from the general formula for the estimation of the integral parameter can significantly enhance the method's information content.

According to the authors of the referenced works [14–28], testing on a four-ball tribometer provides highly reproducible results across multiple experiments and offers valuable data. In our view, conducting tests on a four-ball tribometer should be the primary and essential step in the study of lubricants, both liquid and plastic. These tests help to identify areas for further research in terms of developing promising lubricants and determining the initial optimal content of additives and dopants in the base lubricant [31].

A prerequisite for such studies, constituting the second stage of testing, is the implementation of bench tests on actual machinery or controlled operation tests. During this phase, optimal concentrations will be refined, and factors such as the friction losses and wear rates of the primary tribo-systems in the machinery will be determined when employing new lubricants. For example, in works [32,33], plant-based additives were examined as environmentally friendly additives to lubricants using a four-ball tribometer. Studies involving nanomaterial-based additives, such as fullerenes [34,35], were conducted in works [34–40]. Such investigations necessitate rapid and efficient laboratory tests to identify directions for further research. The work [28] demonstrates the utilization of an integral parameter in assessing the tribological properties of lubricants in mathematical models.

An analysis of the methods using a four-ball tribometer, as outlined in ASTM D4172 and ASTM D2783, reveals several crucial indicators.

Point B on the ASTM D2783 characteristic curve, which characterizes a lubricant, determines whether the lubricant possesses anti-wear properties. These properties are typically conferred by surfactants and their physical adsorption to the friction surface. A shift in point B to the right on the graph indicates the presence of a surfactant in the lubricant, functioning as an anti-wear additive. The primary mechanism of interaction between such additives and the friction surface is physical adsorption.

Conversely, the ASTM D2783 standard does not mandate tests at point A. For instance, GOST 9490 prescribes test conditions at point A, which entails a load of 196 N or 20 kg. This load applies to all lubricants except gear oils. For gear oils, the load at point A is increased to 392 N or 40 kg. Consequently, an AB segment is established, signifying specific physical attributes. This range represents the performance of anti-wear additives in a lubricant with a broader AB range situated lower on the graph field, indicating the superior anti-wear properties of the lubricant.

Section 2, as presented in the ASTM D2783 standard, is the BC segment, which characterizes the rapid increase in the wear scar on the three lower balls as the load increases. Since the load increases in accordance with the standard's load series, the temperature in the friction zone also rises. The elevated temperature causes the physical adsorption of lubricant molecules to the friction surface to diminish. Consequently, the wear scar and temperature on the friction surface increase rapidly. This rise in temperature indicates chemical reactions between the extreme pressure additives and the friction surface material. On the curve, there is a transition from point B to point C, accompanied by deceleration (or a halt) in the growth of the wear scar. The BC segment signifies the presence and effective operation of extreme pressure additives in the lubricant. If extreme pressure (EP) additives are not available, at point C, the scoring of the friction surfaces begins, and the balls may weld together. However, a large BC segment leads to a high wear rate in the materials due to active chemical processes occurring on the friction surfaces. The seizure of the friction surfaces is eliminated, and the wear rate of the materials is increased. A smaller BC segment indicates superior extreme pressure properties of the lubricant.

Section 3, represented by the CD segment, characterizes the performance of the extreme pressure additives in the lubricant. With an increase in the test load, following the standard's load series, the temperature in the friction zone increases significantly. The temperature is the driving force behind the chemical reactions on the friction surfaces that prevent the scuffing or welding of the balls. The physical meaning of the CD segment indicates the effectiveness of the extreme pressure additives in the lubricant. The wider the CD segment, the better the extreme pressure properties of the lubricant. However, the height of point D on the graph negatively affects the wear resistance. Due to the active reactions of the chemical elements in extreme pressure additives with the metal surface, the size of the wear scar increases significantly. Such a lubricant prevents seizure but results in a high wear rate. Ideally, point D should be positioned as far to the right as possible on the graph field and below.

The next load after point D is the welding load, at which either the four-ball pyramid welds together in an oil bath or the four-ball tribometer automatic shuts down. This occurrence signifies the loss of extreme pressure properties in the lubricant.

In our view, a test on a four-ball tribometer can serve as the initial stage in evaluating liquid and plastic lubricants, offering a preliminary assessment of their tribological characteristics. This method enables the determination of the directions for further research in the development of new lubricants at a minimal cost.

The purpose of this investigation was to establish an integral parameter for the evaluation of the tribological properties of lubricants, providing a differentiated assessment of the contribution of anti-wear and extreme pressure additives or dopants.

2. Materials and Methods

The testing unit, following the four-ball scheme, is depicted in Figure 1. It consists of three fixed lower balls (1) and one upper rotating ball (2) secured with a collet (3). The upper rotating ball (2) is subjected to pressure, represented by the load N, against the three lower balls (1).

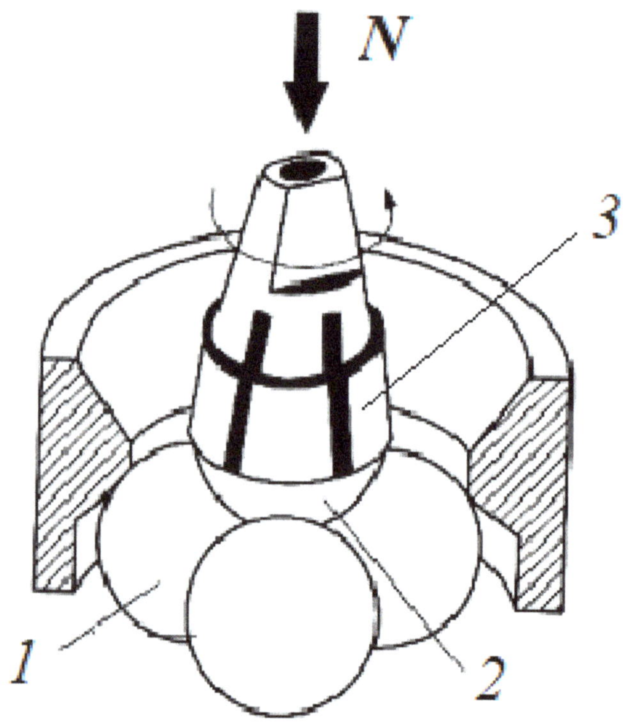

Figure 1. Scheme of a four-ball friction unit. 1—three lower fixed balls; 2—rotating upper ball; 3—collet; *N*—load.

The external view of the four-ball assembly is shown in Figure 2, and a four-ball tribometer is illustrated in Figure 3.

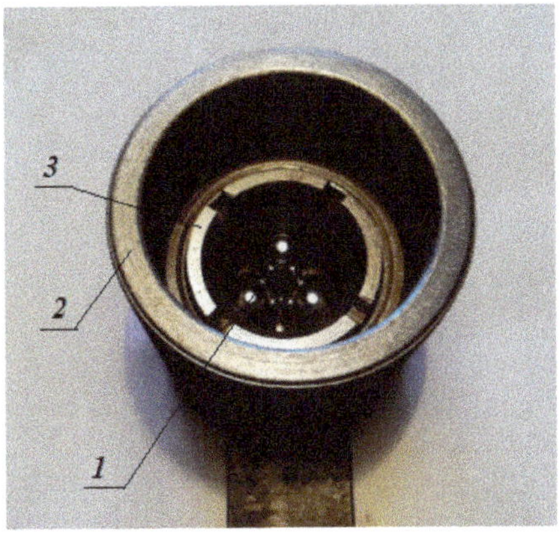

Figure 2. External view of the four-ball assembly. 1—fixed lower balls; 2—a glass for a lubricant to be tested; 3—nut for fixture of the three lower balls.

Figure 3. External view of a four-ball tribometer.

The foundation for the development of an integral parameter for the assessment of the tribological properties of lubricants, with a differentiated evaluation of the contribution of anti-wear and extreme pressure additives, was presented in [27]. The study introduced an energy parameter—the specific wear work of the test material (ball-bearing steel balls) in the tested lubricating medium. This parameter characterizes the amount of friction work required to remove a unit volume of material from the friction surface, expressed as J/m^3, in the form of wear particles. The specific wear work serves as an integral energy parameter whose value can be used to evaluate the anti-wear and extreme pressure properties of lubricants. This manuscript is a continuation of the work presented in [26] and offers an improved technique for the assessment of tribological properties. This work takes into account the requirements outlined in the ASTM D4172 and ASTM D2783 standards.

The specific wear work A is calculated using the following expression:

$$A = \frac{f \cdot P \cdot l}{d^3},\ \mathrm{J/m^3} \tag{1}$$

where f—the coefficient of friction during tests on a four-ball tribometer;

P—the load chosen based on the load range specified in the ASTM D2783 standard, dimension N;

l—the friction path during testing, dimension m;

d—the average diameter of the wear spots on the three lower balls during the test, dimension m.

This parameter can be applied to evaluate the anti-wear and extreme pressure properties of liquid and plastic lubricants.

The methodological approach involves employing this parameter according to Formula (1) for all three ranges, AB, BC, and CD, that characterize the lubricant. Such a methodological approach allows for differentiation in the effectiveness of anti-wear and extreme pressure additives. The sum of these properties results in an integral parameter reflecting the tribological properties of the lubricant [29,30].

As mentioned previously, the AB range is utilized to assess anti-wear properties, and the specific wear work for this range, denoted as A_{AB}, is calculated using the following expression:

$$A_{AB} = \sum_{i=196}^{P_B} \frac{f_i \cdot P_i \cdot l_i}{d_i^3}, \ \mathrm{J/m^3} \tag{2}$$

where $\sum_{i=196}^{P_B}$ —the sum of the values of tests from a load of 196 N (20 kg) to the load preceding the point where the wear scar diameter increases. Such a load is denoted as P_B. According to the aforementioned standards for four-ball tribometer testing, this increase in diameter should be at least 0.1 mm.

f_i—the friction coefficient value at loads ranging from 196 N to P_B.

P_i—the load value from 196 N to P_B, N, applied according to the load range specified in ASTM D2783. Tests must be carried out sequentially at all loads specified in the standard.

l_i—the friction path during testing, with the standard recommending a testing duration of 10 s, resulting in a friction path of l = 5.88 m.

d_i—the average diameter of the wear spots on the three lower balls under loads ranging from 196 N to P_B, N, dimension m.

It should be noted that if the test duration remains consistent at 10 s, then the friction path remains the same, l = 5.88 m.

The higher the value of the specific wear work within the AB segment, the better the anti-wear properties of the lubricant.

The second segment, *BC*, which signifies the shift from anti-wear properties to extreme pressure properties, can be determined using the following expression:

$$A_{BC} = \sum_{P_{B+1}}^{P_C} \frac{f_k \cdot P_k \cdot l_k}{d_k^3}, \ \mathrm{J/m^3} \tag{3}$$

where $\sum_{P_{B+1}}^{P_C}$ —the total value of tests from the load at which the wear scar diameter increases by at least 0.1 mm, denoted as P_{B+1}, up to the load where the increase in the wear scar diameter ceases, denoted as P_C.

f_k—friction coefficient value under loads ranging from P_{B+1} to P_C.

P_k—the load value from P_{B+1}, N, to P_C, N, applied according to the load series specified in ASTM D2783. Tests should be performed sequentially at all loads outlined in the standard.

l_k—the friction path during testing, equal to 5.88 m.

d_k—the average diameter of the wear spots on the three lower balls under loads ranging from P_{B+1}, N, to P_C, N, dimension m.

Based on experience gained from testing various lubricants, it is observed that the range of the BC segment typically encompasses no more than two loads from the load range.

The greater the specific wear work value in the BC segment, the faster the EP additives of the lubricant come into effect, and the more effective their action.

The third segment, denoted as CD, characterizes the performance of extreme pressure additives and can be determined using the following expression:

$$A_{CD} = \sum_{P_{C+1}}^{P_D} \frac{f_j \cdot P_j \cdot l_j}{d_j^3},\ \text{J/m}^3 \tag{4}$$

where $\sum_{P_{C+1}}^{P_D}$ —the total value of tests from the load where the growth in the wear scar ceases, denoted as P_{C+1}, up to the load that precedes the welding load, denoted as P_D.

f_j— the friction coefficient value under loads ranging from P_{C+1} to P_D.

P_j—the load value from P_{C+1}, N, to P_D, N, applied according to the load series specified in ASTM D2783. Tests must be conducted sequentially at all loads outlined in the standard.

l_j—the friction path during testing, equal to 5.88 m.

d_j—the average diameter of the wear spots on the three lower balls under loads ranging from P_{C+1}, N, to P_D, N, dimension m.

Based on the experience of testing various lubricants, as well as additives and dopants used with them, it is observed that not all lubricants under high loads ensure the welding of balls during testing. Some lubricants do not result in welding. Instead, there is accelerated wear of the three lower and one upper balls. In such cases, the standard recommends stopping the test with the load when the wear scar exceeds 3 mm.

The greater the value of the specific wear work in the CD segment, the better the extreme pressure properties of the lubricant. This area can be referred to as the "carrying capacity" of the lubricant, as extreme pressure additives ensure the operation of tribosystems under increased loads and elevated temperatures of the friction surfaces.

With the expressions for the calculation of the specific work of friction on three segments of the curve characterizing the lubricant's performance—see Formulas (2)–(4)—we can derive an expression for the calculation of the integral parameter of the lubricant's tribological properties, represented by the formula

$$A = A_{AB} + A_{BC} + A_{CD},\ \text{J/m}^3 \tag{5}$$

The physical interpretation of this integral parameter for the assessment of the tribological properties of lubricant A, with a differentiated assessment of the contribution from each of these segments, is illustrated in Figure 4.

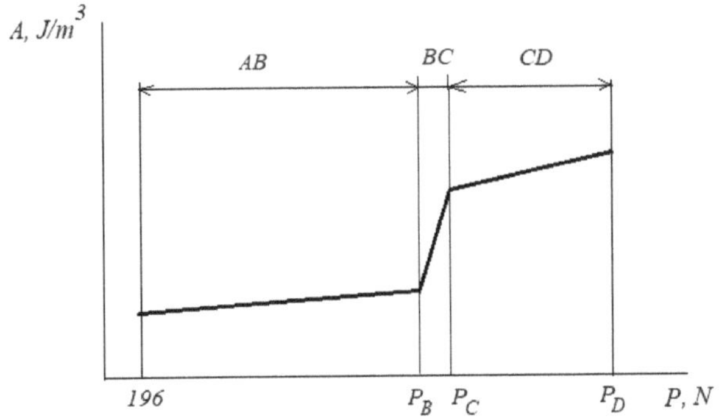

Figure 4. Physical interpretation of the integral parameter for assessment of the tribological properties of lubricants.

The developed methodological approach allows us, within the framework of the ASTM D2783 standard, to determine the presence of antiwear and extreme pressure properties of a lubricant, as well as to differentially evaluate the contribution of each of the three sections to the integral properties.

3. Results of the Experiments

The purpose of the experimental studies was to confirm the information content and reproducibility of the developed methodological approach and the integral parameter in assessing the tribological properties of various lubricants.

For testing, hydraulic fluids (hydraulic oils), engine oils and gear oils were chosen.

Hydraulic fluids and motor and transmission oils belong to the category of liquid lubricants. The purpose of our research was to develop an integral parameter to evaluate the tribological properties of liquid lubricants only. Excluding from this list, for example, hydraulic fluids that simultaneously serve as lubricants for hydraulic units, would make our study incomplete. Therefore, our study examined all types of liquid lubricants.

As hydraulic liquids simultaneously perform the functions of lubricants, we selected working liquids (ISO 6743-4), the tribological properties of which are presented in Table 1. The tribological properties of motor oils are presented in Table 2, and those of transmission oils in Table 3. Each type of lubricant is represented by at least three types. For instance, hydraulic fluids include four types (Table 1), motor oils include seven types (Table 2) and transmission oils include three types (Table 3).

Table 1. Tribological properties of hydraulic oils.

Lubricant Classification	$A_{AB} \times 10^{11}$, J/m^3	$A_{BC} \times 10^{11}$, J/m^3	$A_{CD} \times 10^{11}$, J/m^3	$A \times 10^{11}$, J/m^3
ISO-L-HH	43.54	2.12	0	45.66
ISO-L-HL	57.64	8.87	0.69	67.2
ISO-L-HM	75.45	11.26	1.84	88.59
ISO-L-HV	83.13	13.5	2.3	98.93

Table 2. Tribological properties of motor oils.

Lubricant Classification	$A_{AB} \times 10^{11}$, J/m^3	$A_{BC} \times 10^{11}$, J/m^3	$A_{CD} \times 10^{11}$, J/m^3	$A \times 10^{11}$, J/m^3
SAE 20W-30 API SF	92.93	13.13	2.49	108.55
SAE 10W-30 API SH	113.2	16.12	2.89	132.21
SAE 20W-40 API CD	124.2	17.8	3.1	145.1
SAE 20W-40 API SJ/CF	159.7	19.5	3.72	182.92
SAE 20W-40 API CF-4	198.7	21.7	4.5	225.27
SAE 10W-40 API SL/CF	224.65	23.4	4.87	252.92
SAE 5W-30 APICH	258.26	25.2	5.18	288.64

Table 3. Tribological properties of gear oils.

Lubricant Classification	$A_{AB} \times 10^{11}$, J/m³	$A_{BC} \times 10^{11}$, J/m³	$A_{CD} \times 10^{11}$, J/m³	$A \times 10^{11}$, J/m³
SAE 85W-90 API GL-3	67.55	13.44	8.08	89.07
SAE 80W-90 API GL-4	165.72	24.64	12.43	202.79
SAE 75W-90 API GL-5	213.25	31.75	15.54	260.54

The choice of oil brands was intended to cover the entire range of tribological properties of lubricants, from minimum values to maximum, in accordance with the requirements of the ASTM D4172 and ASTM D2783 standards. With this methodological approach, we considered the entire range of properties that exist today in the liquid lubricant market. We do not refer to specific brands of lubricants but to their classification according to SAE and API.

The abovementioned lubricants were tested in accordance with the ASTM D2783 standard, using the specified load range. Testing for hydraulic and motor oils began with a load of 196 N or 20 kg, with a sequential load series. For gear oils, testing commenced with a load of 392 N or 40 kg.

Based on the results of three repetitions, we calculated the average diameter values of the drift flames on the three lower balls, the root mean square values of the registration values of the diameters of the drift flames during the experimental studies and the coefficient of variation in the measurement of the diameter.

The root mean square deviation of the diameter values of the wear spot during experimental studies was determined using the following formula:

$$S_d = \sqrt{\frac{1}{n}\sum_{i=1}^{n}(d_i - d_{av})^2}, \text{m}, \tag{6}$$

where d_i, d_{av}—the diameter of each wear spot in the sample and the average value for the number of repetitions n, respectively, dimension m.

n—the number of results obtained during the testing of the coating material.

The coefficient of variation in the diameter measurements of wear spots was determined as follows:

$$v_d = \left(\frac{S_d}{d_{av}}\right) \cdot 100\%. \tag{7}$$

The experimental sample of wear spot diameter values obtained during the experimental studies was assessed for uniformity and reproducibility across multiple experiments using Cochran's C test:

$$G_p = \frac{S^2_{d(\max)}}{\sum\limits_{i=1}^{n} S^2_{d(i)}}, \tag{8}$$

where $S^2_{d(\max)}$—the maximum value of the dispersion of the wear spot diameters (d_i) measured during the experiment;

$S^2_{d(i)}$—the value of the variance of the i-th experiment for the parameters to be measured during the experiment—d_i.

The hypothesis was tested using the following condition:

$$G_p < G_{table}, \tag{9}$$

where G_{table}—the tabular value for Cochran's C test, with a predefined confidence factor q = 0.90.

If condition (9) is met, it indicates that the obtained experimental results exhibit uniformity and reproducibility.

4. Discussion

As indicated in the tables above, the integral parameter A—see Formula (5)—dimension J/m^3, exhibits significant variation across different lubricants. For instance, hydraulic oils have A values ranging from 45.66 to 98.93 $\times 10^{11}$ J/m^3, engine oils show values between 108.55 and 288.64 $\times 10^{11}$ J/m^3 and gear oils exhibit values ranging from 89.07 to 260.54 $\times 10^{11}$ J/m^3.

The most pronounced result was obtained when examining the anti-wear and extreme pressure properties separately. The value of the A_{AB} range indicates the presence of anti-wear additives or dopants in the lubricant. Notably, motor oils (as shown in Table 2) tend to display the maximum values of this range, while hydraulic oils (as seen in Table 1) exhibit the minimum values. This range serves as a useful metric for the evaluation of the effectiveness of anti-wear additives and dopants in a lubricant.

On the other hand, the value of the A_{CD} range signifies the presence of extreme pressure additives or dopants within the lubricant. Here, gear oils (Table 3) tend to exhibit the maximum values. For instance, the A_{CD} range varies within 8.08–15.54$\times 10^{11}$ J/m^3. In contrast, hydraulic oils (Table 1) display the minimum values, ranging from 0 to 2.3 $\times 10^{11}$ J/m^3. This range is instrumental in assessing the effectiveness of EP additives and lubricant dopants.

Furthermore, the value of the A_{BC} range provides insights into the effectiveness of the transition from antiwear to extreme pressure properties within the lubricant. Gear oils (Table 3) generally show higher values in the A_{BC} range, signifying a more effective transition to extreme pressure additives. This suggests that the gear oils selected for testing possessed a superior set of extreme pressure additives compared to the engine oils presented.

The laboratory test results for various lubricants assessed on a four-ball tribometer led to the following conclusion (as summarized in Table 4).

1. All test results were reproducible and uniform, with the calculated values of Cochran's C test (Formula (8)) being less than the tabulated values at a given confidence level of probability q = 0.95. This applies to scenarios where the number of test repetitions was 3 and the number of logged values was 9. Consequently, the condition was satisfied; see Formula (9).
2. The coefficient of variation during testing did not exceed 18%; see Formula (7). This indicates the minimal scatter of the test results.
3. The technique presented in this article and the integral parameter for the evaluation of the tribological properties of lubricants—see Formula (5)—can be used at the initial stage of the qualifying laboratory tests of new lubricants, additives, and dopants of various origins.

The physical interpretation of the tribological properties of lubricants for various purposes, along with characteristic areas of operation for various additives, is shown in Figure 5.

The format adopted for the presentation of the research results in Figure 5 allows us to illustrate the physical meaning of the integral parameter used to assess the tribological properties of liquid lubricants. The presented Figure 5 clearly differentiates among various categories of lubricants. The x-axis represents the load range that the lubricant can withstand, characterizing its extreme pressure properties. Greater values indicate better anti-scuff properties. The y-axis reflects the ability of the lubricant to prevent wear, with higher integral parameter A values signifying better anti-wear properties.

The developed technique and the integral parameter for the estimation of the tribological properties of lubricants differ from those that are already known because they are based on an energy approach—the specific work of wear determined on the test material (balls made of ball-bearing steel) in a lubricating medium, as presented in this research. This parameter characterizes the amount of friction work required to remove a unit volume of material from the friction surface (dimension J/m^3) in the form of wear particles. This approach enables the evaluation of the lubricant as a whole and the differentiation of its components, including its anti-wear and extreme pressure properties.

Table 4. Lubricant statistical parameters.

Lubricant Classification	$S_d \times 10^{-3}$, m	v_d, %	G_p	G_{table}
ISO-L-HH	0.155	17.32	0.432	0.478
ISO-L-HL	0.137	17.24	0.413	0.478
ISO-L-HM	0.113	16.26	0.412	0.478
ISO-L-HV	0.09	14.53	0.388	0.478
SAE 20W-30 API SF	0.052	13.23	0.410	0.478
SAE 10W-30 API SH	0.043	12.52	0.410	0.478
SAE 20W-40 API CD	0.041	11.83	0.394	0.478
SAE 20W-40 API SJ/CF	0.04	11.54	0.380	0.478
SAE 20W-40 API CF-4	0.04	11.72	0.345	0.478
SAE 10W-40 API SL/CF	0.039	10.49	0.340	0.478
SAE 5W-30 APICH	0.039	10.28	0.340	0.478
SAE 85W-90 API GL-3	0.055	14.32	0.410	0.478
SAE 80W-90 API GL-4	0.047	13.55	0.380	0.478
SAE 75W-90 API GL-5	0.042	10.45	0.375	0.478

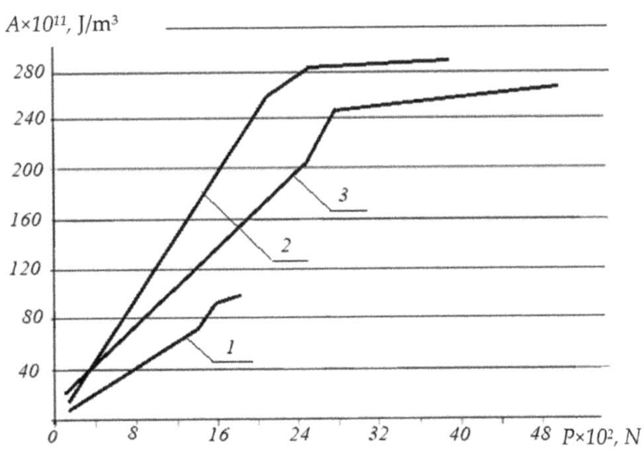

Figure 5. Physical interpretation of the integral parameter for assessment of the tribological properties of lubricants: 1—ISO-L-HM hydraulic oil; 2—engine oil SAE 5W-30, API CH; 3—gear oil SAE 75W-90, API GL-5.

Further development of this method for the testing of lubricants on a four-ball tribometer aims to enhance the efficiency of screening tests during the development of new lubricants, such as vegetable oils. Lubricant additives using nanomaterials, such as fullerenes, are actively being developed, as discussed in [26]. Additionally, this parameter can be utilized in modeling friction and wear processes in tribo-systems to monitor the lubricating medium, as presented in [24,27].

5. Conclusions

Based on the analysis of standards for the testing of lubricants—both liquid and plastic—an energy parameter is proposed: the specific work of wear of the test material (steel balls) in the lubricating medium to be investigated. Unlike known parameters, this energy parameter is associated with the work of friction expended in removing the volume of material from the friction surface during testing. Experimental evidence demonstrates that this proposed parameter evaluates the range of operation of anti-wear and extreme pressure additives in the lubricant, allowing for the differentiation of their effectiveness.

The results of laboratory tests on various lubricants using a four-ball tribometer verify that, according to the developed method, all tests are reproducible and uniform, as confirmed by Cochran's C test. The coefficient of variation during testing does not exceed 18%, indicating minimal variation in the test results.

The technique presented in this article and the integral parameter for the assessment of the tribological properties of lubricants can be applied in the initial stages of laboratory selection tests for new lubricants, additives, and dopants of various origins.

Author Contributions: Conceptualization, T.M.A.A.-Q., F.A., J.H. and V.V.; methodology, T.M.A.A.-Q., O.M., A.V., F.A., A.K. (Andrey Kravtsov) and J.H.; software, V.V., O.M., A.K. (Andrey Kravtsov) and F.A.; validation, O.M. and A.K. (Andrii Kondratiev); formal analysis, T.M.A.A.-Q. and V.V.; investigation, F.A., O.M. and A.V.; resources, T.M.A.A.-Q., O.M., A.V. and A.K. (Andrey Kravtsov); data curation, J.H., O.M. and A.V.; writing—original draft preparation, T.M.A.A.-Q., F.A., J.H., A.K. (Andrii Kondratiev). and V.V.; writing—review and editing, T.M.A.A.-Q., O.M., A.K. (Andrii Kondratiev) and A.V.; visualization, F.A., J.H., A.K. (Andrey Kravtsov) and O.M.; supervision, V.V., O.M. and A.V. All authors have read and agreed to the published version of the manuscript.

Funding: This research received no external funding.

Institutional Review Board Statement: Not applicable.

Informed Consent Statement: Not applicable.

Data Availability Statement: The data presented in this study are available on request from the corresponding author.

Conflicts of Interest: The authors declare no conflict of interest.

References

1. Stachowiak, G.W.; Batchelor, A.W.; Stachowiak, G.B. 3—Tribometers. *Tribol. Ser.* **2004**, *44*, 25–78. [CrossRef]
2. Ji, X.; Chen, Y.; Zhao, G.; Wang, X.; Liu, W. Tribological Properties of $CaCO_3$ Nanoparticles as an Additive in Lithium Grease. *Tribol. Lett.* **2011**, *41*, 113–119. [CrossRef]
3. Joysula, S.K.; Dube, A.; Patro, D.; Veeregowda, D.H. On the Fictitious Grease Lubrication Performance in a Four-Ball Tester. *Lubricants* **2021**, *9*, 33. [CrossRef]
4. Beheshti, A.; Huang, Y.; Ohno, K.; Blakey, I.; Stokes, J.R. Improving tribological properties of oil-based lubricants using hybrid colloidal additives. *Tribol. Int.* **2020**, *144*, 106130. [CrossRef]
5. Yadav, G.; Tiwari, S.; Jain, M.L. Tribological analysis of extreme pressure and anti-wear properties of engine lubricating oil using four ball tester. *Miner. Process.* **2018**, *5*, 248–253. [CrossRef]
6. Xie, H.; Wei, Y.; Jiang, B.; Tang, C.; Nie, C. Tribological properties of carbon nanotube/SiO_2 combinations as water-based lubricant additives for magnesium alloy. *J. Mater. Res. Technol.* **2021**, *12*, 138–149. [CrossRef]
7. Kumar, N.; Saini, V.; Bijwe, J. Tribological investigations of nano and micro-sized graphite particles as an additive in lithium-based grease. *Tribol. Lett.* **2020**, *44*, 124. [CrossRef]
8. Akhtar, K.; Khalid, H.; Haq, I.U.; Malik, A. Improvement in tribological properties of lubricating grease with quartz-enriched rice husk ash. *Tribol. Int.* **2016**, *93*, 58–62. [CrossRef]
9. Saini, V.; Bijwe, J.; Seth, S.; Ramakumar, S. Potential exploration of nano-talc particles for enhancing the anti-wear and extreme pressure performance of oil. *Tribol. Int.* **2020**, *151*, 106452. [CrossRef]

10. Kumar, N.; Saini, V.; Bijwe, J. Exploration of Talc nanoparticles to enhance the performance of Lithium grease. *Tribol. Int.* **2021**, *162*, 107107. [CrossRef]
11. Havrylenko, Y.; Kholodniak, Y.; Halko, S.; Vershkov, O.; Miroshnyk, O.; Suprun, O.; Dereza, O.; Shchur, T.; Śrutek, M. Representation of a Monotone Curve by a Contour with Regular Change in Curvature. *Entropy* **2021**, *23*, 923. [CrossRef] [PubMed]
12. Khasawneh, A.; Qawaqzeh, M.; Miroshnyk, O.; Danylchenko, D.; Minakova, K.; Potryvai, A. Methodology for Accounting for the Influence of Dust Cover on the Performance of a Photovoltaic System for Matlab. In Proceedings of the 20th IEEE International Conference on Modern Electrical and Energy Systems, MEES 2021, Kremenchuk, Ukraine, 21–24 September 2021. [CrossRef]
13. Fox, N.J.; Tyrer, B.; Stachowiak, G.W. Boundary Lubrication Performance of Free Fatty Acids in Sunflower Oil. *Tribol. Lett.* **2004**, *16*, 275–281. [CrossRef]
14. Zhao, F.Y.; Li, G.T.; Zhang, G.; Wang, T.M.; Wang, Q.H. Hybrid effect of ZnS sub-micrometer particles and reinforcing fibers on tribological performance of polyimide under oil lubrication conditions. *Wear* **2017**, *380–381*, 86–95. [CrossRef]
15. Rawat, S.S.; Harsha, A.P.; Deepak, A.P. Tribological performance of paraffin grease with silica nanoparticles as an additive. *Appl. Nanosci.* **2019**, *9*, 305–315. [CrossRef]
16. Havrylenko, Y.; Kholodniak, Y.; Halko, S.; Vershkov, O.; Bondarenko, L.; Suprun, O.; Miroshnyk, O.; Shchur, T.; Śrutek, M.; Gackowska, M. Interpolation with Specified Error of a Point Series Belonging to a Monotone Curve. *Entropy* **2021**, *23*, 493. [CrossRef]
17. Rawat, S.S.; Harsha, A.P.; Das, S.; Deepak, A.P. Effect of CuO and ZnO nano-additives on the tribological performance of paraffin oil–based lithium grease. *Tribol. Trans.* **2020**, *63*, 90–100. [CrossRef]
18. Rawat, S.S.; Harsha, A.P.; Khatri, O.P. Tribological investigations of two-dimensional nanostructured lamellar materials as additives to castor-oil-derived lithium grease. *J. Tribol.* **2022**, *144*, 091902. [CrossRef]
19. Vyavhare, K.; Aswath, P.B. Tribological Properties of Novel Multi-Walled Carbon Nanotubes and Phosphorus Containing Ionic Liquid Hybrids in Grease. *Front. Mech. Eng.* **2019**, *5*, 15. [CrossRef]
20. Iegorov, O.; Iegorova, O.; Miroshnyk, O.; Savchenko, O. Improving the accuracy of determining the parameters of induction motors in transient starting modes. *Energetika* **2020**, *66*, 15–23. [CrossRef]
21. Saini, V.; Bijwe, J.; Seth, S.; Ramakumar, S. Augmenting the lubrication performance of nano-oils through synergistic co-functioning of nanoparticles. *Tribol. Int.* **2023**, *182*, 108332. [CrossRef]
22. Kondratiev, A.; Píštěk, V.; Smovziuk, L.; Shevtsova, M.; Fomina, A.; Kučera, P.; Prokop, A. Effects of the temperature–time regime of curing of composite patch on repair process efficiency. *Polymers* **2021**, *13*, 4342. [CrossRef]
23. Saini, V.; Bijwe, J.; Seth, S.; Ramakumar, S. Role of base oils in developing extreme pressure lubricants by exploring nano-PTFE particles. *Tribol. Int.* **2020**, *143*, 106071. [CrossRef]
24. Panickar, R.; Sobhan, C.B.; Chakravorti, S. Investigations on tribological properties of non-catalytic CVD synthesized carbon spheres in lubricant. *Diam. Relat. Mater.* **2020**, *106*, 107834. [CrossRef]
25. Saini, V.; Bijwe, J.; Seth, S.; Ramakumar, S. Interfacial interaction of PTFE sub-micron particles in oil with steel surfaces as excellent extreme-pressure additive. *J. Mol. Liq.* **2021**, *325*, 115238. [CrossRef]
26. Solanki, R.; Saini, V.; Bijwe, J. Exploration of PTFE sub-micron particles for enhancing the performance of commercial oils. *Surf. Topogr. Metrol. Prop.* **2021**, *9*, 025005. [CrossRef]
27. Vojtov, V.; Biekirov, A.; Voitov, A. The quality of the tribosystem as a factor of wear resistance. *Int. J. Eng. Technol.* **2018**, *7*, 25–29. [CrossRef]
28. Al-Quraan, T.M.A.; Alfaqs, F.; Alrefo, I.F.S.; Vojtov, V.; Voitov, A.; Kravtsov, A.; Miroshnyk, O.; Kondratiev, A.; Kučera, P.; Píštěk, V. Methodological Approach in the Simulation of the Robustness Boundaries of Tribosystems under the Conditions of Boundary Lubrication. *Lubricants* **2023**, *11*, 17. [CrossRef]
29. Al-Quraan, T.M.A. Influence Evaluation of the RVS Friction Geomodifier on Tribotechnical Parameters of the Contact in Non-Stationary working Conditions. *Tribol. Ind.* **2020**, *42*, 121–130. [CrossRef]
30. Al-Quraan, T.M.A. A Manifestation of Non Hydrodynamic Factors in Steady State Conditions of Friction. *J. Mech. Eng.* **2017**, *14*, 109–124. Available online: https://ir.uitm.edu.my/id/eprint/21124 (accessed on 31 December 2017).
31. Kondratiev, A.V.; Kovalenko, V.O. Optimization of design parameters of the main composite fairing of the launch vehicle under simultaneous force and thermal loading. *Space Sci. Technol.* **2019**, *25*, 3–21. [CrossRef]
32. Ali, A.I.; Hamid, M.K.A.; Abas, M.A.B.; Said, M.F.M.; Mazali, I.I.; Daud, Z.H.C.; Opia, A.C. Experimental analysis of tribological performance of base ficus carica vegetable oil with different polymers as additives using four ball tribometer. *Mater. Today Proc.* 2023, *in press*. [CrossRef]
33. Kumar, A.; Chaudhary, R.; Singh, R.C. Tribological performance of various blends of commercial SAE 40 oil and novel apricot oil-based bio-lubricant using a four-ball tester tribometer. *Int. J. Mater. Prod. Technol.* **2023**, *67*, 166–177. [CrossRef]
34. Tarun, S.; Keishing, S.; Sharma, S. Tribological analysis of nanoparticles in lubricants using four ball tester. *Int. Res. J. Mod. Eng. Technol. Sci.* **2023**, *5*, 5064–5066. [CrossRef]
35. Anurag, S.; Prashant, C.; Mamatha, T.G. A review on tribological performance of lubricants with nanoparticles additives. *Mater. Today Proc.* **2020**, *25*, 586–591. [CrossRef]
36. Shahnazar, S.; Bagheri, S.; Hamid, S. Enhancing lubricant properties by nanoparticle additives. *Int. J. Hydrog. Energy* **2015**, *41*, 3153–3170. [CrossRef]

37. Ali, I.; Basheer, A.; Kucherova, A.; Memetov, N.; Pasko, T.; Ovchinnikov, K.; Pershin, V.; Tkachev, A. Advances in carbon nanomaterials as lubricants modifiers. *J. Mol. Liq.* **2019**, *279*, 251–266. [CrossRef]
38. Yao, Y.; Wang, X.; Guo, J.; Yang, X.; Xu, B. Tribological property of onion-like fullerenes as lubricant additive. *Mater. Lett.* **2008**, *62*, 2524–2527. [CrossRef]
39. Rapoport, L.; Feldman, Y.; Homyonfer, M.; Cohen, H.; Sloan, J.; Hutchison, J.L.; Tenne, R. Inorganic fullerene-like material as additives to lubricants: Structure–function relationship. *Wear* **1999**, *225–229*, 975–982. [CrossRef]
40. Yunusov, F.A.; Breki, A.D.; Vasilyeva, E.S.; Tolochko, O.V. The Influence of Nano Additives on Tribological Properties of Lubricant Oil. *Mater. Today Proc.* **2020**, *30*, 632–634. [CrossRef]

lubricants

MDPI

Article

Methodological Approach in the Simulation of the Robustness Boundaries of Tribosystems under the Conditions of Boundary Lubrication

Tareq M. A. Al-Quraan [1,*], Fadi Alfaqs [2], Ibrahim F. S. Alrefo [1], Viktor Vojtov [3], Anton Voitov [3], Andrey Kravtsov [3], Oleksandr Miroshnyk [4], Andrii Kondratiev [5], Pavel Kučera [6] and Václav Píštěk [6,*]

1 Department of Technical Sciences, Ma'an College, Al-Balqa Applied University, Al-Salt 19117, Jordan
2 Department Mechanical Engineering, Faculty of Engineering Technology, Al-Balqa Applied University, Amman 11134, Jordan
3 Department of Transport Technologies and Logistics, State Biotechnological University, 61052 Kharkiv, Ukraine
4 Department of Electricity Supply and Energy Management, State Biotechnological University, 61052 Kharkiv, Ukraine
5 Department of Materials Science and Engineering of Composite Structures, O.M. Beketov National University of Urban Economy in Kharkiv, Marshal Bazhanov Str. 17, 61002 Kharkiv, Ukraine
6 Institute of Automotive Engineering, Brno University of Technology, Technická 2896/2, 616-69 Brno, Czech Republic
* Correspondence: tarjorggg@bau.edu.jo (T.M.A.A.-Q.); pistek.v@fme.vutbr.cz (V.P.); Tel.: +420-541-142-271 (V.P.)

Abstract: In the presented work, a methodical approach was developed for determining rational operation modes of tribosystems, taking into account their design. This approach makes it possible in the designing stage, according to the predicted operating modes, to calculate the limits and margins of stable work in operation. The definition of the robustness of the tribosystem and the criteria for assessing the robustness are formulated based on the theory of stability of technical systems. It is shown that such a methodical approach allows for determining the modes of the rational operation of the designed structures without damaging the friction surfaces. Experimental studies have proven that not all designs of tribosystems lose stability due to the appearance of friction surface burrs. There are designs where the loss of stability occurs upon the appearance of accelerated wear. The developed criteria take into account two options for the loss of stability. An experimental verification of the modes of loss of stability of tribosystems was performed by the appearance of a burr or the beginning of accelerated wear with the calculated values of the robustness criteria. The obtained results allow us to conclude that the modeling error is within 8.3–18.7%, which is a satisfactory result in the study of friction and wear processes. Robustness criteria is based on the coefficient of friction RR_f and wear rate RR_I, and must be used when designing new constructions of tribosystems. Theoretical calculations of such criteria and the dependence of their change on changing the predicted operating modes will allow for justifying rational operating modes within their stability.

Keywords: tribosystem; limit lubrication mode; burr of friction surfaces; accelerated wear; stability of technical systems; robustness; modelling; robustness criteria; modelling error; modes of operation

1. Introduction

The stable operation of tribosystems in the entire load–speed range of operation is the most important characteristic that the reliability indicators of machines and mechanisms depend on. The range of the stable operation of tribosystems is predicted at the stage of the construction development of new machines and is based on the experimental data of previous designs or experimental data of laboratory bench tests.

The most promising direction is mathematical modelling of the limits of the stable operation (robustness) of tribosystems, which will allow for significantly reducing material resources during the construction of new equipment. The development of such models is based on the theoretical foundations of the stability of technical systems, developed by O.M. Lyapunov, who created the modern theory of the stability of motion of mechanical systems determined by a finite number of parameters.

The concept of the stability of tribosystems is the most important qualitative assessment of their dynamic properties and depends on constructive, technological, and operational factors. Through the stability of the operation of tribosystem, it is possible to restore the original mode of operation after the removal of an external influence. In addition, an important parameter is the limit of the loss of the stable operation of the tribosystem, i.e., the amount of load and the sliding speed when burr or accelerated wear occurs [1,2]. According to this, three types of tribosystems were formulated in [1].

1. Stable tribosystems, which, after being brought out of equilibrium by any external disturbance (change in load or sliding speed, or the short-term cessation of lubricant supply), after the removal of this disturbance, return to the original stable state, i.e., the established operating mode.
2. Neutral tribosystems, which, after removing the disturbance, switch to a state of stable operation in a new mode, which is different from the original one.
3. Unstable tribosystems, which, as they are brought out of equilibrium by any external disturbance, do not return to the original stable state after the removal of this disturbance, but switch to the mode of accelerated wear or to burr, i.e., cease operation.

It should be noted that majority of studies are devoted to the failure of tribosystems due to the burring (seizing) of friction surfaces. However, there are options for terminating the operation of tribosystems due to the appearance of accelerated wear.

The task of determining the limits of the stable operation (robustness) of tribosystems, considering the design, manufacturing technology, and the load–speed range of operation, is an urgent task. This task is solved in the stage of designing new structures of tribocouplers, which allows for increasing their resource and reliability in operation. This article deals only with sliding friction processes.

An analysis of recent publications on the use of mathematical models for predicting the reliability and service life of tribosystems at the stage of development and design allows us to conclude that entropy analysis is promising in the study of energy dissipation during friction [3–6]. For example, in [4], a mathematical model for predicting the reliability of rolling bearings is developed on the basis of the hierarchical Bayesian method of maximum entropy. Theoretical developments that show the relationship of wear with the thermodynamic concept of entropy are presented in [5–7]. This approach allowed the authors of the work to study the effect of consistent grease on the resources of tribosystems.

On the basis of the provisions of the thermodynamics of open systems in [8–10], equations for the conservation of mass and energy are proposed, which allow for simulating various modes of energy dissipation in a tribosystem, considering the friction surface. These equations can be considered as a tool for modelling various modes of energy dissipation in the dynamic contact of friction surfaces with the development of practical recommendations for the design of new structures.

For simulating the work of tribosystems and predicting the resource, the authors of [11–13] used the achievements of artificial intelligence, particularly artificial neural networks. The authors showed that the use of this approach allowed for modelling and forecasting the wear and reliability of tribosystems during exploitation.

The authors of [10–14] paid attention to the importance of considering the friction surface in mathematical models. For example, the importance of considering the roughness of the friction surface in the models is noted in [14], the values of the actual contact area are noted in [15,16], the change in the roughness of friction surfaces during running-in or running-out in is noted [17], and the distribution of actual pressures and stresses over the contact surface is noted in [18,19]. According to the results of the authors, consideration of the listed factors will increase the accuracy of modelling and predicting the resources during the operation of tribosystems.

The authors of [20,21] note that mathematical models of the functioning of tribosystems are tools for designing new friction units. The work presents a methodical approach and models that consider the design of the friction unit, which increases the accuracy of forecasting.

The authors of [22,23] aimed their work at considering the stability and efficiency of tribosystems. The models created by the authors were built on the basis of stochastic models of surface roughness. It was shown that the tribological behavior in contact depended on the properties of the friction surfaces and the spectrum of loads. Moreover, edge effects at the boundaries of the contact area affected the pressure distribution and membrane thickness, which should be taken into account while modelling.

The authors of [24–26] noted that when developing mathematical models for the functioning of tribosystems, it is necessary to consider wear during the running-in period. Because of the intense deformation of microprotrusions, the running-in mode allows for the formation of an equilibrium roughness of friction surfaces and a change in the structure of thin surface layers, which are analyzed in [24]. Optimal running-in conditions will help increase the service life and reliability of tribosystems.

The analyses of publications determining the boundaries of the stable operation of tribosystems shows that these are the operating modes when the scuffing or seizing of friction surfaces occurs with the transfer of material from one triboelement to another.

The difficulty in predicting the failure of tribosystems is a fundamental problem of tribology, which is noted in [27]. The article presents an analysis of various factors (residual stresses, surface free energy, and surface topography) that lead to the initial phase of the process of burr. The studies presented in [28] are aimed at a better understanding of the occurrence of burr. The authors affirm that the parametric assessment of the state of the friction surface is of crucial importance in the technology of preventing burrs.

The durability limits of the tribosystem and the transition from stable to unstable operating conditions, namely seizing, were studied in [29]. The purpose of the research was to determine the boundaries of the stable operation of tribosystems and to use the obtained results while designing tribosystems.

The authors of [30] studied the influence of random overloads on the occurrence of burrs in tribosystems. The results of the research were the determination of the boundary of the exit of the tribosystem to the burr. The mechanism of occurrence of burrs, based on adiabatic shear instability, was experimentally evaluated in [31]. The authors of the work related the mechanism of burr formation to the density of dislocations in the surface layer.

The equations for modeling the burr of friction surfaces, which were obtained on the basis of adiabatic shear instability, are presented in [32]. With the help of equations, it is proposed to model the mechanism of loss of strength from the surface layer of the material during adiabatic shear. The main factor in the loss of strength is the work of friction. On the basis of this mechanism, a quantitative criterion for predicting reliability has been developed, where burr formation is the result of competition between thermal strength loss and strain hardening during adiabatic shear.

The mechanism of burr formation in friction nodes was the subject of tribological studies in [33]. According to the authors, the approach to this problem requires taking into account a large number of working parameters for predicting burr. This work proposes using the parameter (pressure × velocity × time) instead of the frequently used parameter (pressure × velocity). According to the authors, such an approach will allow for taking into account the duration or frequency of loads during the operation of friction nodes. These parameters can significantly affect the stability of the properties of the lubricating membrane, especially in the context of the formation of an elastohydrodynamic membrane or boundary layer.

The results of the gear tests on a test bench with changes in the torque, sliding speed, oil bath temperature, gear geometry, and viscosity of the base lubricant are presented in [34]. The developed model takes into account normal pressure and tangential stresses at several points along the line of the clutch of the gear. Gear seizure results are analyzed using two different approaches: one takes into account the general parameters of the gear, including the clutch line, and the other, based on the local parameters of the roughness of the friction surfaces, is determined using the mixed lubrication model. An analysis of the roughness level was used to explain the causes of burr formation.

A method for predicting the appearance of burr on friction surfaces, using the analysis of the dynamics of the transient process, is proposed in [35]. Quasi-static and nonlinear dynamic models take into account the increase in temperature during the transition process when the load changes during operation. The model allows for determining the thickness of the lubricating membrane on the friction surface and for determining the limit of the burr when the load changes.

The processes of the stability of tribosystems when oscillations occur in frictional contact were studied in [36,37]. For example, strict criteria for the stable operation of tribosystems in various modes of frictional sliding and the occurrence of vibrations were presented in [36]. The stability of the tribosystem, with the help of a complex analysis of the oscillating elements of the tribosystem and their influence on the process was investigated in [37].

Studies on the burrs of the gear were presented in [38,39]. The flash point on the actual contact patches was explored in [38]. The optimization of the design of the gearbox according to the criterion of power loss due to friction in the gears was described in [39].

The analysis of works devoted to the determination of the limits of the stable operation of tribosystems allowed us to conclude that when developing and justifying such criteria, it is necessary to take into account the constructive, technological, and operational factors of tribosystems. In the works listed above, the geometric dimensions of tribocouplers, physical and mechanical properties of combined materials of triboelements, the tribological properties of the lubricating medium, and theroughness of friction surfaces are not sufficiently taken into account. Accounting for the mentioned factors will allow us to extend the obtained criteria to a wide class of tribosystems and make such an analysis systematic.

2. Materials and Methods

The purpose of this study is to develop a methodical approach for determining the robustness of tribosystems in conditions of extreme lubrication, modelling changes in the robustness of tribosystems at the stage of constructive development, and experimental verification of the modelling error.

To substantiate the methodical approach in the research, we will use the equation of the dynamics of tribosystem functioning, which is given in [1].

The friction unit is modeled as a series and parallel connection of dynamic links from which the equivalent transfer function of the friction unit is obtained. On the basis of the equivalent transfer function, a differential equation for the process of operation of the friction unit is written.

The differential equation of the third order is written in operator form, as follows:

$$(T_1T_2T_3)p^3 + (T_1T_2 + T_1T_3 + T_2T_3)p^2 + (T_1 + T_2 + T_3 + K_2K_3T_1)p + K_2K_3 + 1 = (K_1K_2T_3)p + K_1K_2, \quad (1)$$

where p is the differentiation operator, which is equivalent to writing d/dt;

T_1, T_2, T_3 are the time constants in the dimension of seconds (s);

K_1, K_2, K_3 are the gain coefficients, which are dimensionless quantities.

In [1], it was proven that time constant T_1 characterizes the change in the structure of the materials of the surface layers during run-in. It is defined by the following expression:

$$T_1 = \frac{t_{run}}{3}, \quad (2)$$

where t_{run} ia the tribosystem run-in time in the dimension of seconds.

Time constant T_2 has a physical sense of the time, during which the temperature is equalized by the volumes of triboelement materials when the load and sliding speed change. It is defined by the following expression:

$$T_2 = \frac{254 \cdot V_{run}}{a_{run} \cdot d_{acs} \cdot n_{acs}}, \quad (3)$$

where V_{run} is the stated volume of materials of the moving and stationary triboelements of the tribosystem, in the dimension of m^3, and is determined by Expressions (2) and (3), which are given in [40], taking into account the total volume of the material of the friction node;

a_{run} is the given coefficient of the thermal conductivity of the materials of the moving a_r and stationary a_s triboelements, in the dimension of m^2/s, and is calculated according to Expression (6), which is given in [40];

d_{acs} is the diameter of the actual contact spot, in the dimension of m, and it is calculated according to the expression given in [41];

n_{acs} is the number of contact spots on the friction surface, and is calculated according to the expression given in [41].

Time constant T_3 characterizes the time until the friction and wear parameters stabilize in the new operating mode of the tribosystem. Time constant T_3 has different expressions to the model wear rates $T_{3(I)}$ and friction coefficient $T_{3(f)}$:

$$T_{3(I)} = \frac{428000 \cdot V_{def}}{\dot{\varepsilon}_{run} \cdot d_{acs}^3 \cdot n_{acs}}, \quad (4)$$

$$T_{3(f)} = \frac{1068000 \cdot V_{def}}{\dot{\varepsilon}_{run} \cdot d_{acs}^3 \cdot n_{acs}}, \quad (5)$$

where V_{def} is the given volume of deformable surface layers, measured in m^3, and is determined by the expression given in [40], using Formulas (15) and (16);

$\dot{\varepsilon}_{run}$ is the indicated value of the rate of deformation of the surface layers of the materials of the movable and immovable triboelements, in the dimension of c^{-1}, which is calculated according to the formulas given in [30], using Formulas (8)–(10).

The coefficient K_1 includes the degree of external influence on the tribosystem, considering the design features, which was proven in [40]:

$$K_1 = \frac{Q_{max}}{Q_0}, \quad (6)$$

where Q_0 and Q_{max} are the initial value of the Q-factor of the tribosystem and the value of the Q-factor that was formed during the run-in. They are determined by the formulas given in [41].

Coefficient K_2 characterizes the sensitivity of the tribosystem to changes in the external conditions:

$$K_2 = \frac{5500 \cdot W_{TR} \cdot K_f}{Q_{max} \bullet a_{run}}, \quad (7)$$

where W_{TR} is the speed of dissipation in the tribosystem, measured in J/s, which is calculated according to the equations given in [1], using Formula (8);

K_f is the shape coefficient of the tribosystem, in the dimension of m^{-1}, and includes the friction areas and the volumes located under the friction areas of the moving and stationary triboelements. It is calculated according to the equation given in [41].

Coefficient K_3 characterizes the ability of the tribosystem to change the structure of the surface layers of the triboelement materials during transient processes:

$$K_3 = \frac{170 \cdot RS_{TS(\max)}^2 \cdot a_{run}}{\dot{\varepsilon}_{run}}, \quad (8)$$

where $RS_{TS(max)}$ is the maximum value of the rheological properties of the combined materials in the tribosystem after running-in, measured in dimension m^{-1}, and is calculated according to the equations given in [42].

Formulas (7) and (8) are obtained for the specific design of the friction unit. The design is taken into account by using the shape factor, K_f. The shape factor takes into account the magnitude of the friction areas and the volumes of material that are placed under the working areas of friction. Values of 5500 and 170 were introduced to improve the accuracy of the simulation. When changing the structure of the friction unit, it is necessary to adjust the values of these coefficients.

To analyze the stability of the technical systems, several special methods have been developed, which are called stability criteria in the theory of automatic regulation. Stability criteria are divided into two varieties—algebraic and frequency.

Algebraic criteria are analytical and frequency criteria are grapho analytical [43–45]. At the same time, all criteria are based on the theory of the stability of technical systems developed by Lyapunov [46–49].

The algebraic criterion, Hurwitz's criterion [50–52], is the most common criterion and is used to determine the stability of technical systems when the characteristic equation is known. As the characteristic equation acts the left part of the differential equation in operator form (1).

Let us write down the characteristic equation, making a substitution and equating it to zero:

$$A_0p^3 + A_1p^2 + A_2p + A_3 = 0, \tag{9}$$

where

$$A_0 = T_1 \cdot T_2 \cdot T_3, \tag{10}$$

$$A_1 = T_1 \cdot T_2 + T_1 \cdot T_3 + T_2 \cdot T_3, \tag{11}$$

$$A_2 = T_1 + T_2 + T_3 + K_2K_3T_1, \tag{12}$$

$$A_3 = K_2 \cdot K_3 + 1, \tag{13}$$

According to Hurwitz's algebraic stability criterion, a technical system is stable when all coefficients A_i of the characteristic Equation (9) are more than zero. This is a necessary condition for stability:

$$A_0 > 0; A_1 > 0; A_2 > 0; A_3 > 0, \tag{14}$$

A sufficient condition for stability is that all determinants from the coefficients A_i of the characteristics in Equation (9) are more than zero.

If at least one of the determinants is equal to zero, then the system is on the verge of loss of stability. If at least one of the determinants is negative, then the system is unstable. In the tribosystem, burr or accelerated wear occur.

According to the rules, let us write down all of the determinants for the characteristic Equation (9):

$$\Delta_1 = A_1 > 0, \tag{15}$$

$$\Delta_2 = \begin{vmatrix} A_1 & A_3 \\ A_0 & A_2 \end{vmatrix} = A_1 \cdot A_2 - A_0 \cdot A_3 > 0, \tag{16}$$

$$\Delta_3 = \begin{vmatrix} A_1 & A_3 & 0 \\ A_0 & A_2 & 0 \\ 0 & A_1 & A_3 \end{vmatrix} = \begin{vmatrix} A_1 & A_3 \\ A_0 & A_2 \end{vmatrix} A_3 > 0 = (A_1 \cdot A_2 - A_0 \cdot A_3) \cdot A_3 > 0, \tag{17}$$

Let us write down the value of the determinantΔ_3 in the following form:

$$\Delta_3 = (A_1 \cdot A_2 - A_0 \cdot A_3) \cdot A_3 > 0, \tag{18}$$

Expression (18) is a necessary and sufficient condition for the stable functioning of the tribosystem according to the Hurwitz criterion. Using this expression, you can determine the range of stable operation for the tribosystem, namely the range of robustness. The more the value of Δ_3, measured in dimension s_3, the more the margin of robustness for the tribosystem. When the value $\Delta_3 = 0$, the tribosystem works on the verge of loss of stability, with negative values for Δ_3, and there is burr or accelerated wear and the tribosystem has "lost stability".

To compare tribosystems and build a rating on the margin of stable operation, it is necessary to obtain a dimensionless parameter—a criterion that depends on constructive, technological, and operational factors. The most acceptable, in our opinion, is the analysis of the expression of the stability of technical systems based on the Hurwitz algebraic criterion, using formula (18).

Based on the formulated approach, let us write down, in a general form, the evaluation criterion for the margin of stable operation of the tribosystem—the robustness criterion:

$$RR = \frac{A_1 \cdot A_2}{A_0 \cdot A_3} > 1, \tag{19}$$

where RR is criterion of tribosystem robustness, with a dimensionless value.

The physical meaning for determining the robustness of the technical systems is given in [53].

An analysis of Equation (1) shows that processes of friction and wear in the tribosystem, especially the running-in processes, depend on the first derivative of the input effect on the tribosystem, i.e., on the load speed. The load speed of the tribosystem can be considered by the coefficient of the change in the load speed, which is proportional to the right-hand side of the differential Equation (1):

$$k_d \approx (K_1 K_2 T_3)\frac{dW_i}{dt_l} + K_1 K_2, \tag{20}$$

where the amount of load (external influence during experimental studies) on the tribosystem is determined by the following expression:

$$W_i = N \cdot v_{sl}, \tag{21}$$

where N is the load on the tribosystem, N;

v_{sl} is the sliding speed, m/s;

t_l is the load change time, s.

With the help of laboratory experimental studies carried out on various designs of tribosystems with different amounts of load on the tribosystem and different rates of load change, an expression for calculating the coefficient of change of the load speed was obtained.

Determining the robustness of the tribosystem using the friction coefficient parameter is completed as follows:

$$k_{d(f)} = 0.62 \ln\left(\frac{K_1 \cdot K_2 \cdot T_{3(f)}}{t_l}\right). \tag{22}$$

Evaluating the operation of the tribosystem by the wear rate parameter is completed as follows:

$$k_{d(I)} = 0.95 \ln\left(\frac{K_1 \cdot K_2 \cdot T_{3(I)}}{t_l}\right). \tag{23}$$

Including the expressions of the coefficient of change in the load speed (22) and (23), which were obtained according to the results of experimental studies, equations for determining the robustness of tribosystems (19), including expressions (10)–(13), will be presented in the following form.

To determine the robustness of the tribosystem by the friction coefficient:

$$RR_f = \frac{\left(\left(T_1T_2 + T_1T_{3,f} + T_2T_{3,f}\right) \times \left(T_1 + T_2 + T_{3,f} + K_2K_3T_1\right)\right)}{\left(T_1T_2T_{3,f}K_2K_3 + T_1T_2T_{3,f}\right) \cdot k_{d(f)}} > 1. \quad (24)$$

To determine the robustness of the tribosystem by the rate of wear:

$$RR_I = \frac{((T_1T_2 + T_1T_{3,I} + T_2T_{3,I}) \times (T_1 + T_2 + T_{3,I} + K_2K_3T_1))}{(T_1T_2T_{3,I}K_2K_3 + T_1T_2T_{3,I}) \cdot k_{d(I)}} > 1. \quad (25)$$

It is necessary to calculate criteria for the robustness of the tribosystem RR_f and RR_I for each load of the operating range of tribosystems, including the analysis of the obtained values. If the value of the criterion is more than one, then the tribosystem works in a stable range. The more the value of the robustness criterion, the more the margin of sustainable work.

If the value of the criterion is equal to one, the tribosystem works on the verge of loss of stability. If the value of the criterion is less than one, the tribosystem has lost stability, and burr or accelerated wear has occurred.

To answer the question of which the parameter loss of stability occurred in, it is necessary to calculate two criteria: according to the coefficient of friction, formula (24) and according to the rate of wear, formula (25). The value of the criterion, which first becomes less than one, answers the question on which parameter the loss of stability occurred.

Figure 1 presents the theoretical dependence of the change in the robustness criterion of various structures of tribosystems according to the friction coefficient on the value of the input influence, using formula (24). Input influence on the tribosystem, *W*, W, is represented as the product of load, *N*, and sliding speed, m/s, using formula (21). Different designs of tribosystems are included by the value of the *Q*-factor, Q_o, J/m^3, [41,42].

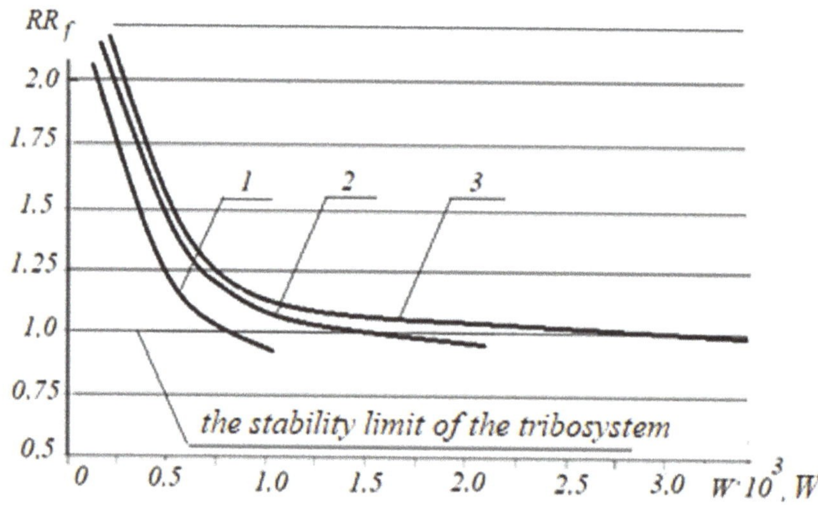

Figure 1. Dependencies of changes in the robustness criterion of various designs of tribosystems according to the coefficient of friction on the value of the input impact: 1—steel 5140 + steel 5140, K_f = 6.25 m^{-1}, hydraulic oil HH, ISO-L-HL, Q_0 = 1.12·10^{10} J/m^3; 2—steel 5140 + bronze C61900, K_f = 12.5 m^{-1}, motor oil SAE 40, APICC, Q_0 = 5.5·10^{10} J/m^3; 3—steel 5140 + brass CW723R, K_f = 14.5 m^{-1}; transmission oil SAE120, APIGL-4, Q_0 = 7.69·10^{10} J/m^3.

Experimental studies were carried out on a universal friction machine according to the «ring-ring» kinematic scheme. The design of the friction machine is presented in [41].

The friction unit is represented by movable sample 1 and fixed sample 2. The friction between the samples occurs over the area of the ends. During the tests, fixed sample 2 is pressed against movable sample 1 by load N. Friction occurs over the area of the ends. The movable sample performs a rotational movement around its axis. An acoustic emission sensor GT300 is mounted on fixed sample 2. The acoustic emission signal from the GT300 transducer with a frequency of 100–800 kHz is transmitted to a preamplifier and then to a USB oscilloscope, which acts as an A/D converter. Then, the signal in a digital code is entered into the computer to calculate the diagnostic parameters.

The AE signal from the friction zone is recorded by a broadband sensor GT300 (100–800 kHz), Figure 2, which is installed on a fixed triboelement, transmitted to an amplifier, then, in analog form, to a PV6501 USB oscilloscope, which performs the functions of an analog-to-digital converter and spectrum frequency analyzer at the same time. After processing in a USB oscilloscope, the digital signal enters the computer, where it is processed by special software.

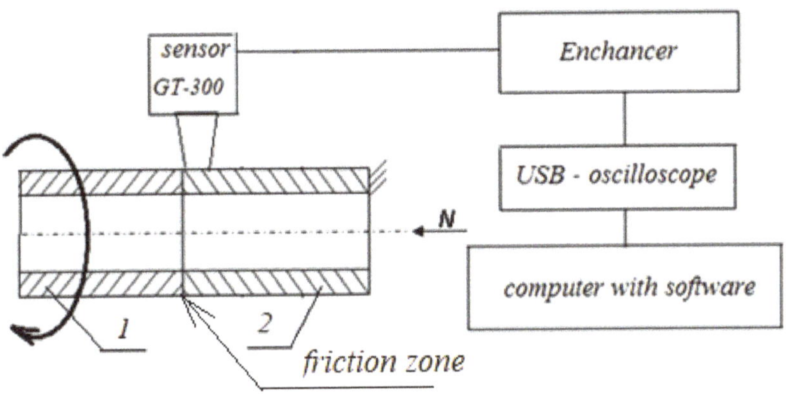

Figure 2. Block diagram of experimental equipment for recording and processing AE signals: 1—movable triboelement; 2—fixed triboelement; *N*—load.

Samples for testing are presented in Figure 3, where movable sample 1 is made of steel5140, and fixed sample 2 is made of various materials: steel 5140, bronze C61900, and brass CW723R. Thus, during the tests, it was possible to obtain various constructions of tribosystems, for example, a movable sample of steel 5140 + fixed sample of steel 5140; movable sample steel 5140 + fixed sample bronze C61900; movable sample steel 5140 + fixed sample brass CW723R. At the same time, the friction area of fixed samples 2 was changed, which made it possible to obtain different coefficients for the shape of the tested tribosystems.

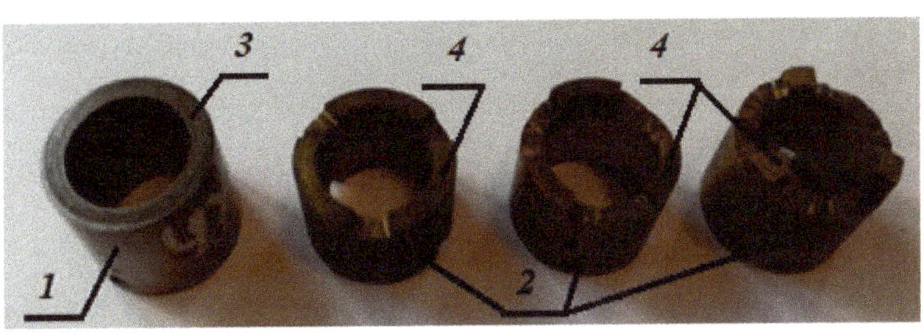

Figure 3. Samples for testing: 1—movable sample steel 5140; 2—fixed samples: steel 5140, bronze C61900, brass CW723R with different friction areas on the end surfaces; 3—friction surfaces of the moving sample; 4—friction surfaces of a fixed sample.

During the tests, the load on the tribosystem increased at different load speeds: 1 s, 10 s, and 20 s.

The loss of stability of tribosystems, according to the coefficient of friction was determined by using the value of the moment of friction, which was registered by the recorder of machine 2070 SMT-1. According to the parameter of the start of accelerated wear, the loss of stability was determined by using of the acoustic emission method. The measuring complex and the method for recording the acoustic emission signals are given in [54]. To record acoustic eradiation from the friction zone, the acoustic emission sensor is installed on a stationary triboelement.

According to the results of three repetitions, the average burr load or the onset of accelerated wear, the root mean square deviation of values of registered values during experimental studies, formula (26); coefficient of variation of measurements of external influence, when the occasion of loss of stability of the tribosystem occurs, formula (27); modelling error according to the formula (28), were determined.

The root-mean-square deviation of the values of the external influence during experimental research is represented by the formula:

$$S_{W_b} = \sqrt{\frac{1}{n}\sum_{i=1}^{n}\left(W_{b(i)} - W_{b(aw)}\right)^2}, \tag{26}$$

where $W_{b(i)}$, $W_{b(aw)}$ are the values of external influence on the tribosystem, formula (21), at which there is a loss of stability (burr or accelerated wear), which is measured during the experiment and is averaged over the number of repetitions n.

The coefficient of variation of measurements of external influence, when the occasion of loss of stability of the tribosystem occurs, is determined by the expression:

$$v_{W_b} = \left(\frac{S_{W_b}}{W_{b(aw)}}\right) \cdot 100\%, \tag{27}$$

The relative error of modelling the robustness of tribosystems is determined by the expression:

$$e_{W_b} = \left|\frac{W_{b(\exp)} - W_{b(s)}}{W_{b(\exp)}}\right| \cdot 100\%, \tag{28}$$

where $W_{b(exp)}$, $W_{b(s)}$ is the value of the magnitude of the external influence on the tribosystem at which a loss of stability (burr or accelerated wear) occurs, which is measured during the experiment and according to the results of the simulation.

3. The Results of the Experimental Research

The results of modelling the robustness of tribosystems and the results of the experimental testing are presented in Table 1. The purpose of the experimental research is to compare the calculated values of the magnitude of the external influence on the tribosystem $W_{b(s)}$, at which a loss of stability occurs (burr or accelerated wear), according to formulas (24) and (25), when RR_f and RR_I are equal to the unit, with the results of the experiment $W_{b(exp)}$.

The nature of the change in the value of the external influence on the tribosystem in which the loss of stability $W_{b(exp)}$ for various designs of tribosystems occurs, which are evaluated by the shape factor K_f, m^{-1}, is presented in the first block of Table 1. Experimental studies of the limits of the stable operation of tribosystems are performed for the tribosystem steel 5140 + bronze C61900, and lubricating medium E_y= 3.6·10^{14} J/m^3, (motor oil SAE 40, API CC). The parameter E_y takes into account the tribological properties of the lubricating medium. The technique for determining E_y is described in [41]. The sliding speed is not changed and is equal v_{sl} = 0.5 m/s. During the experiment, the values of the shape factor of the tribosystem vary, K_f = 6.25–22.6 m^{-1}. This value is obtained by changing the friction area of the stationary triboelement. The load is increased from the initial value of 400 N, to the value when the tribosystem loses stability, i.e., burr or accelerated wear occur. The initial roughness of the friction surfaces for all experimental conditions was Ra = 0.2 μm.

Table 1. The results of testing the error of modelling the range of robustness of various designs of tribosystems.

Construction of the Tribosystem	*Wb(s)*, N·m/s	*Wb(exp)*, N·m/s	*SWb*, N·m/s	*vWb*, %	*eWb*, %
Steel 5140 + bronze C61900, motor oil SAE 40, APICC, $E_y = 3.6 \cdot 10^{14}$ J/m^3, K_f = 6.25 m^{-1}	1300 *(I)* *	1200 *(I)* *	200	16.6	8.3
Steel 5140 + bronze C61900, motor oil SAE 40, APICC, $E_y = 3.6 \cdot 10^{14}$ J/m^3, K_f = 12.5 m^{-1}	1900 *(I, f)* *	1700 *(I, f)* *	300	17.6	11.7
Steel 5140 + bronze C61900, motor oil SAE 40, APICC, $E_y = 3.6 \cdot 10^{14}$ J/m^3, K_f = 22.6 m^{-1}	2300 *(f)* *	2000 *(f)* *	400	20.0	15.0
Steel 5140 + bronze C61900, K_f = 12.5 m^{-1}, hydraulic oil HH, ISO-L-HL, ($E_y = 2.43 \cdot 10^{14}$ J/m^3)	850 *(f)* *	1000 *(f)* *	200	20.0	15.0
Steel 5140 + bronze C61900, K_f = 12.5 m^{-1}, motor oil SAE 40, APICC, ($E_y = 3.6 \cdot 10^{14}$ J/m^3)	1900 *(I, f)* *	1700 *(I, f)* *	300	17.6	11.7
Steel 5140 + bronze C61900, K_f = 12.5 m^{-1}, transmission oil SAE 120, APIGL-4, ($E_y = 4.18 \cdot 10^{14}$ J/m^3)	2600 *(I)* *	2350 *(I)* *	350	14.8	10.6
Steel 5140 + steel5140, ($RS_{TS(max)}$ = 326.7 m^{-1}), K_f = 12.5 m^{-1}, motor oil SAE 40, APICC, $E_y = 3.6 \cdot 10^{14}$ J/m^3	1300 *(f)* *	1150 *(f)* *	250	21.7	13.0
Steel 5140 + bronze C61900, ($RS_{TS(max)}$ = 436.0 m^{-1}), K_f = 12.5 m^{-1}, motor oil SAE 40, APICC, $E_y = 3.6 \cdot 10^{14}$ J/m^3	1900 *(I, f)* *	1700 *(I, f)* *	300	17.6	11.7
Steel 5140 +brass CW723R, ($RS_{TS(max)}$ = 460.9 m^{-1}), K_f = 12.5 m^{-1}, motor oil SAE 40, APICC, $E_y = 3.6 \cdot 10^{14}$ J/m^3	1950 *(I, f)* *	1800 *(I, f)* *	300	16.6	8.3
Tribosystem No. 1: Steel 5140 + Steel 5140, ($RS_{TS(max)}$ = 326.7 m^{-1}), K_f = 6.25 m^{-1}, hydraulic oil HH, ISO-L-HL, ($E_y = 2.43 \cdot 10^{14}$ J/m^3), $Q_0 = 1.12 \cdot 10^{10}$ J/m^3	650 *(f)* *	800 *(f)* *	150	18.7	18.7
Tribosystem No. 2: Steel 5140 + bronze C61900, ($RS_{TS(max)}$ = 436.0 m^{-1}), K_f = 12.5 m^{-1}, motor oil SAE 40, APICC, $E_y = 3.6 \cdot 10^{14}$ J/m^3, $Q_0 = 5.5 \cdot 10^{10}$ J/m^3	2000 *(I, f)* *	1700 *(I, f)* *	300	17.6	17.6
Tribosystem No. 3: Steel 5140 + brass CW723R, ($RS_{TS(max)}$ = 460.9 m^{-1}), K_f = 14.5 m^{-1},transmission oil SAE 120, APIGL-4, $E_y = 4.18 \cdot 10^{14}$ J/m^3, $Q_0 = 7.69 \cdot 10^{10}$ J/m^3	3100 *(I)* *	2700 *(I)* *	350	12.9	14.8

Notation in the table: *(f)* *—loss of stability according to the coefficient of friction (burr); *(I)* *—loss of stability due to the appearance of accelerated wear; *(I, f)* *—loss of stability due to the coefficient of friction (burr) or the appearance of accelerated wear.

With K_f = 6.25 m^{-1} the loss of stability for the tribosystem occurs after the appearance of a burr. When increasing the form factor to K_f = 22.6 m^{-1}, the loss of stability of the tribosystem occurs after the appearance of accelerated wear.

The calculation of the error for determining the limit of the stable operation of tribosystems according to formula (28) when the shape factor of tribosystems changes allows us to confirm that the value of the error is equal to e_W = 8.3–15.0%, at the coefficient of variation v_W = 16.6–20.0%. As it follows from the obtained results, an increase in the shape factor leads to an increase in the modelling error.

The results of experimental studies on the change in the value of the external influence on the tribosystem in which the loss of stability occurs when the tribological properties of the lubricating medium are changed E_y is presented in the second block of Table 1.

The results are presented for the tribosystem: steel 5140 + bronze C61900; form factor K_f = 12.5 m^{-1}. As a factor, the following changes have been selected: hydraulic oil HH, ISO-L-HL, ($E_y = 2.43 \cdot 10^{14}$ J/m^3); motor oil SAE 40, API CC, ($E_y = 3.6 \cdot 10^{14}$ J/m^3); transmission oil SAE 120, API GL-4, (E_y = 4.18·10^{14} J/m^3). The sliding speed did not change and was equal v_{sl} = 0.5 m/s.

The calculation of the error for determining the limit of the stable operation of tribosystems, when the tribological properties of the lubricating medium change, allows us to assert that the value of the modelling error is within e_W = 10.6–15.0%, at the coefficient of variation v_W = 14.8–20.0%. A greater error is inherent in the use of lubricants with low values of tribological properties.

The results of experimental studies on the change in the value of the external influence on the tribosystem in which the loss of stability occurs when the rheological properties of the structure of the combined materials in the tribosystem are changed $RS_{TS(max)}$ are presented in the third block of the Table 1.

The results are presented for the tribosystem: form factor K_f = 12.5 m^{-1}. Lubricating medium—motor oil SAE 40, API CC, (E_y = 3.6·10^{14} J/m^3). As a factor that changes, have been selected: steel 5140 + steel 5140, ($RS_{TS(max)}$ = 326.7 m^{-1}); steel 5140 + bronze C61900, ($RS_{TS(max)}$ = 436.0 m^{-1}); steel 5140 + brass CW723R, ($RS_{TS(max)}$ = 460.9 m^{-1}).

Comparing the experimental results with the results of modelling when changing the rheological properties of the combined materials in the tribosystem allows us to affirm that the error value is within e_W = 8.3–13.0%, at a coefficient of variation v_W = 16.6–21.7%. A greater error is inherent in the use of composite materials with low values of rheological properties.

The results of experimental studies regarding the change in the value of the external influence on the tribosystem in which a loss of stability occurs when all of the above factors are changed, which can be considered by the *Q*-factor of the tribosystem Q_0, are presented in the fourth block of Table 1. The determination of the *Q*-factor of the tribosystem is given in [41].

The results are presented for three tribosystems.

1. Tribosystem №1: «steel 5140 + steel 5140», ($RS_{TS(max)}$ = 326.7 m^{-1}); K_f = 6.25 m^{-1}; lubricating medium E_y = 2.43·10^{14} J/m^3, (HH, ISO-L-HL). The value of the *Q*–factor of the tribosystem Q_0 = 1.12·10^{10} J/m^3.
2. Tribosystem №2: «steel 5140 + bronze C61900», ($RS_{TS(max)}$ = 436.0 m^{-1}); K_f = 12.5 m^{-1}; lubricating medium E_y = 3.6·10^{14} J/m^3, (SAE 40, API CC). The value of the *Q*–factor of the tribosystem Q_0 = 5.5·10^{10} J/m^3.
3. Tribosystem №3: «steel 5140 + brass CW723R», ($RS_{TS(max)}$ = 460.9 m^{-1}); K_f = 14.5 m^{-1}; lubricating medium E_y = 4.18·10^{14} J/m^3, (SAE 120, API GL-4). The value of the *Q*–factor of the tribosystem Q_0 = 7.69·10^{10} J/m^3.

An increase in the *Q*-factor value of the tribosystem, which is subject to change at the same time, impacts: the shape factor, tribological properties of the lubricating medium, rheological properties of the combined materials in the tribosystem, the thermal conductivity of the materials of the moving and stationary triboelements, the load, and the sliding speed. The efficiency of tribosystems varies from values Q_0 = 1.12·10^{10} J/m^3, (tribosystem №1), to values Q_0 = 7.69·10^{10} J/m^3, (tribosystem №3). An increase in the *Q*-factor contributes to an increase in the robustness of tribosystems.

Comparing the experimental results with the simulation results when all of the above factors are changed, makes it possible to assert that the value of the simulation error is within e_W = 14.8–18.7%, at the coefficient of variation v_W = 12.9–18.7%. A bigger error is inherent in the application of tribosystems with low *Q*-values.

The results of the experimental studies allow us to state that not all tribosystems lose stability due to the appearance of burrs on the friction surfaces (according to the friction coefficient parameter), the mark in the table *(f)* *. There are a few possible options, when the loss of stability occurs due to accelerated wear of materials, a mark in the table *(I)* *. Or the appearance of a burr and accelerated wear have equal value, a mark in the table *(I, f)* *.

The experimental data, shown in Table 1, were obtained with a load time for the tribosystem equal to 20 s. At this value for the load time, the coefficient of change for the load speed corresponds to the minimum values: $k_{d(f)} = 3.42$; $k_{d(I)} = 4.37$, formulas (22) and (23). When the load time is reduced, to 1 s, the coefficients increase significantly, for example: $k_{d(f)} = 6.4$; $k_{d(I)} = 7.38$. The introduction of the coefficient for the change in the load speed (22) and (23) into the calculation expressions for determining the robustness of tribosystems (24) and (25) allows for reducing the modelling error.

4. Discussion

According to the obtained research results, it is possible to formulate a definition of the term that characterizes the limits for the stable operation of the tribosystem—the robustness of the tribosystem. This is a dimensionless value that corresponds to the modes of operation of the tribosystem without damaging the friction surfaces.

Calculation of tribosystem robustness criteria RR_f, formula (24), and RR_I, formula (25), makes it possible to determine the rational modes of operation of tribosystems at the stage of design and engineering development. The operation of tribosystems in a certain range will increase their resource and reliability.

The robustness criteria are obtained on the basis of the theory of stability of technical systems, the foundations of which were developed by Lyapunov and can be applied to a diverse class of tribosystem designs. Experimental studies have proven that not all designs of tribosystems lose their stability due to the appearance of burrs on the friction surfaces. There are structures where the loss of stability occurs after the appearance of accelerated wear. Developed criteria RR_f and RR_I consider two options for the loss of stability.

The obtained results for modelling the stable operation of tribosystems and the values of the reserve of stable operation should be used in the design of new structures. By solving constructive, technological, and operational tasks, it is possible to obtain a rational design of the tribosystem, which will satisfy the design tasks and will be operated in a range of speeds and loads with a margin of robustness.

5. Conclusions

A methodical approach has been developed to determine the rational modes of operation of tribosystems, considering their design. This approach makes it possible to calculate the limits and margin of stable operation in exploitation at the design stage, according to the predicted operating modes.

The definition of the robustness of the tribosystem and the criteria for assessing the robustness are formulated on the basis of the theory of stability of technical systems. It is shown that such a methodical approach allows for determining the modes of rational operation of the designed structures without damaging the friction surfaces. Experimental studies have proven that not all designs of tribosystems lose their stability due to the appearance of burrs on the friction surfaces. There are structures where loss of stability occurs after the appearance of accelerated wear. The developed criteria consider two variants of stability loss.

An experimental verification of the modes of loss of stability of tribosystems by the appearance of a burr or the beginning of accelerated wear with the calculated values of the robustness criteria was performed. The obtained results allow us to conclude that the modelling error was within the limits of 8.3–18.7%, which is a satisfactory result in the study of friction and wear processes.

Robustness criteria based on the coefficient of friction RR_f and wear rate RR_I, must be used when projecting new designs of tribosystems. Theoretical calculations of such criteria and the dependence of their change on changing the predicted operating modes will allow to justify rational operating modes within the limits of their stability.

Author Contributions: Conceptualization, T.M.A.A.-Q., F.A., I.F.S.A. and V.V.; methodology, T.A., O.M., A.V., A.K. (Andrey Kravtsov), F.A. and I.F.S.A.; software, V.V., O.M., A.K. (Andrii Kondratiev) and A.V.; validation, A.K. (Andrey Kravtsov), O.M., F.A., P.K. and V.P.; formal analysis, T.M.A.A.-Q., I.F.S.A. and V.V.; investigation, F.A., O.M. and A.K. (Andrey Kravtsov); resources, T.A, O.M., A.V. and A.K. (Andrey Kravtsov); data curation, I.F.S.A., O.M., A.K. (Andrii Kondratiev) and A.V.; writing—original draft preparation, T.M.A.A.-Q., F.A., I.F.S.A. and V.V.; writing—review and editing, T.M.A.A.-Q., F.A., I.F.S.A., A.V., P.K. and V.P.; visualization, F.A., I.F.S.A., O.M. and A.K. (Andrey Kravtsov); supervision, V.V., O.M. and A.V.; funding acquisition, F.A., O.M. and A.K. (Andrey Kravtsov). All authors have read and agreed to the published version of the manuscript.

Funding: The authors gratefully acknowledge the funding from the specific research on BUT FSI-S-20-6267.

Data Availability Statement: Not applicable.

Acknowledgments: The authors thank Brno University of Technology for support.

Conflicts of Interest: The authors declare no conflict of interest.

Nomenclature

T_1, T_2, T_3	time constants, dimension s.
K_1, K_2, K_3	amplification factors, dimensionless values.
t_{run}	tribosystem run-in time, dimension s.
V_{run}	the given volume of materials of the moving and stationary triboelements of the tribosystem, dimension m^3
a_{run}	given coefficient of thermal conductivity of materials of movable a_p and fixed a_n triboelements, dimension m^2/s.
d_{acs}	diameter of the actual spot of contact, dimension m.
n_{acs}	the number of contact spots on the friction surface
V_{def}	given volume of deformable surface layers, dimension m^3.
$\dot{\varepsilon}_{run}$	the magnitude of the rate of deformation of the surface layers of the materials of the movable and fixed triboelements, dimension s^{-1}.
Q_0 and Q_{max}	the initial Q-factor value of the tribosystem and the Q-factor value after run-in, dimension J/m^3.
W_{TR}	speed of dissipation in the tribosystem, dimension J/s.
K_f	tribosystem form factor, dimension m^{-1}.
$RS_{TC(max)}$	the maximum value of the rheological properties of the combined materials in the tribosystem after the completion of running-in, dimension m^{-1}.
N	load on the tribosystem, dimension N.
v_{sl}	sliding speed, dimension m/s.
t_l	load change time, dimension s.
$k_{d(f)}$	coefficient of change of load speed according to the friction coefficient parameter, dimensionless value.
$k_{d(I)}$	coefficient of change of the load speed according to the parameter of the wear rate, dimensionless value.
RR_f	robustness of the tribosystem according to the friction coefficient, dimensionless value.
RR_I	robustness of the tribosystem according to the rate of wear, dimensionless value.
$W_{b(i)}$, $W_{b(aw)}$	the value of the magnitude of the external influence on the tribosystem at which occurs the loss of stability (burr or accelerated wear), according to the results of the experiment and according to the results of simulation, dimension J/s.
E_y	tribological properties of the lubricating medium, dimension J/m^3.
S_{Wb}	root mean square deviation of values of external influence during experimental studies, dimension J/s.
v_{Wb}	coefficient of variation of measurements of external influence, at which the event of loss of stability of the tribosystem occurs, %.
e_{Wb}	relative error of modeling the robustness of tribosystems, %.
R_a	initial roughness of friction surfaces, dimension μm.

References

1. Voitov, A. Structural identification of the mathematical model of the functioning of tribosystems under conditions of boundary lubrication. *Probl. Tribol.* **2021**, *26*, 26–33. Available online: http://tribology.khnu.km.ua/index.php/ProbTrib/article/view/794/1269 (accessed on 8 November 2022). [CrossRef]
2. Kondratiev, A.V.; Gaidachuk, V.E. Mathematical analysis of technological parameters for producing superfine prepregs by flattening carbon fibers. *Mech. Compos. Mater.* **2021**, *57*, 91–100. [CrossRef]
3. Li, H.; Liu, Y.; Liao, H.; Liang, Z. Accelerated Wear Test Design Based on Dissipation Wear Model Entropy Analysis under Mixed Lubrication. *Lubricants* **2022**, *10*, 71. [CrossRef]
4. Ye, L.; Hu, Y.; Deng, S.; Zhang, W.; Cui, Y.; Xu, J. A Novel Model for Evaluating the Operation Performance Status of Rolling Bearings Based on Hierarchical Maximum Entropy Bayesian Method. *Lubricants* **2022**, *10*, 97. [CrossRef]
5. Al-Quraan, T.M.A.; Mikosyanchik, O.O.; Mnatsakanov, R.G. Temperature resistance of the boundary lubrication layers under rolling with slippage condition. *Int. J. Mech. Eng. Appl.* **2017**, *5*, 78–86. [CrossRef]
6. Gurt, A.; Khonsari, M. The Use of Entropy in Modeling the Mechanical Degradation of Grease. *Lubricants* **2019**, *7*, 82. [CrossRef]
7. Bogdanovych, O.I.; Al-Quraan, T.M.A.; Tokaruk, V.V.; Haddad, J.S. Algorithm for tribokinetic modeling tests of triboconjuc-tion materials for industrial products. *Tribol. Ind.* **2020**, *43*, 159–166. [CrossRef]
8. Karl, D. Mass and Energy Balance of a Three-Body Tribosystem. *Lubricants* **2022**, *10*, 95. [CrossRef]
9. Kondratiev, A.; Píštěk, V.; Smovziuk, L.; Shevtsova, M.; Fomina, A.; Kučera, P. Stress–strain behaviour of reparable composite panel with step–variable thickness. *Polymers* **2021**, *13*, 3830. [CrossRef]
10. Al-Quraan, T.M.A.; Mikosyanchik, O.O.; Mnatsakanov, R.G.; Zaporozhets, O.I. Structural-Energy characteristics of tribotechnical contact in unsteady operational modes. *Mod. Mech. Eng.* **2016**, *6*, 91–97. [CrossRef]
11. Rosenkranz, A.; Marian, M.; Profito, F.J.; Aragon, N.; Shah, R. The Use of Artificial Intelligence in Tribology—A Perspective. *Lubricants* **2021**, *9*, 2. [CrossRef]
12. Miroshnuk, O.O.; Tymchuk, S.O. Uniform distribution of loads in the electric system 0.38/0.22 KV using genetic algorithms. *Tech. Electrodyn.* **2013**, *4*, 67–73. Available online: http://www.scopus.com/inward/record.url?eid=2-s2.0-84885913005&partnerID=MN8TOARS (accessed on 8 November 2022).
13. Argatov, I. Artificial Neural Networks (ANNs) as a Novel Modeling Technique in Tribology. *Front. Mech. Eng.* **2019**, *5*, 1074. [CrossRef]
14. Joerger, A.; Reichert, S.; Wittig, C.; Aghdam, D.S.; Albers, A. An Approach for the Transfer of Real Surfaces in Finite Element Simulations. *Lubricants* **2021**, *9*, 77. [CrossRef]
15. Wang, C.; Schipper, D.J. On an Elastoplastic Sliding Model for a Coated Single Asperity. *Lubricants* **2018**, *6*, 96. [CrossRef]
16. Dmitriev, A.I.; Österle, W.; Wetzel, B.; Zhang, G. Mesoscale modeling of the mechanical and tribological behavior of a poly-mer matrix composite based on epoxy and 6 vol.% silica nanoparticles. *Comput. Mater. Sci.* **2015**, *110*, 204–214. [CrossRef]
17. Winkler, A.; Marian, M.; Tremmel, S.; Wartzack, S. Numerical Modeling of Wear in a Thrust Roller Bearing under Mixed Elastohydrodynamic Lubrication. *Lubricants* **2020**, *8*, 58. [CrossRef]
18. Terwey, J.T.; Fourati, M.A.; Pape, F.; Poll, G. Energy-Based Modelling of Adhesive Wear in the Mixed Lubrication Regime. *Lubricants* **2020**, *8*, 16. [CrossRef]
19. Jacq, C.; Nelias, D.; Lormand, G.; Girodin, D. Development of a three-dimensional semi-analytical elastic-plastic contact code. *J. Tribol.* **2002**, *124*, 653–667. [CrossRef]
20. Delprete, C.; Gastaldi, C.; Giorio, L. A Minimal Input Engine Friction Model for Power Loss Prediction. *Lubricants* **2022**, *10*, 94. [CrossRef]
21. Wang, Z.; Yu, H.; Wang, Q. Analytical solutions for elastic fields caused by eigenstrains in two joined and perfectly bonded half-spaces and related problems. *Int. J. Plast.* **2016**, *76*, 1–28. [CrossRef]
22. Orgeldinger, C.; Tremmel, S. Understanding Friction in Cam–Tappet Contacts—An Application-Oriented Time-Dependent Simulation Approach Considering Surface Asperities and Edge Effects. *Lubricants* **2021**, *9*, 106. [CrossRef]
23. Yang, L.; Wang, D.; Guo, Y.; Liu, S. Tribological behaviors of quartz sand particles for hydraulic fracturing. *Tribol. Int.* **2016**, *102*, 485–496. [CrossRef]
24. Scherge, M. The Running-in of Lubricated Metal-Metal Contacts—A Review on Ultra-Low Wear Systems. *Lubricants* **2018**, *6*, 54. [CrossRef]
25. Vojtov, V.A.; Biekirov, A.S.; Voitov, A.V.; Tsymbal, B.M. Running-in procedures and performance tests for tribosystems. *J. Frict. Wear* **2019**, *40*, 376–383. [CrossRef]
26. Kondratiev, A.; Píštěk, V.; Smovziuk, L.; Shevtsova, M.; Fomina, A.; Kučera, P.; Prokop, A. Effects of the Temperature–Time Regime of Curing of Composite Patch on Repair Process Efficiency. *Polymers* **2021**, *13*, 4342. [CrossRef]
27. Wojciechowski, Ł.; Mathia, T.G. Proposal of invariant precursors for boundary lubricated scuffing. *Wear* **2015**, *340–341*, 53–62. [CrossRef]
28. Wojciechowski, Ł.; Wieczorowski, M.; Mathia, T.G. Transition from the boundary lubrication to scuffing—The role of metallic surfaces morphology. *Wear* **2017**, *392–393*, 39–49. [CrossRef]
29. Pusterhofer, M.; Summer, F.; Gódor, I.; Grün, F. Cumulative damage assessment of tribological durability limits. *Wear* **2020**, *456–457*, 203318. [CrossRef]

30. Savolainen, M.; Lehtovaara, A. An experimental approach for investigating scuffing initiation due to overload cycles with a twin-disc test device. *Tribol. Int.* **2017**, *109*, 311–318. [CrossRef]
31. Hershberger, J.; Ajayi, O.O.; Zhang, J.; Yoon, H.; Fenske, G.R. Evidence of scuffing initiation by adiabatic shear instability. *Wear* **2005**, *258*, 1471–1478. [CrossRef]
32. Zhang, C.; Peng, B.; Gu, L.; Wang, T.; Wang, L. A scuffing criterion of steels based on the friction-induced adiabatic shear instability. *Tribol. Int.* **2020**, *148*, 120–134. [CrossRef]
33. Wojciechowski, L.; Mathia, T.G. Focus on the concept of pressure-velocity-time (pVt) limits for boundary lubricated scuffing. *Wear* **2018**, *402–403*, 179–186. [CrossRef]
34. Castro, J.; Sottomayor, A.; Seabra, J. Experimental and analystical scuffing criteria for FZG gears. *Tribol. Ser.* **2003**, *43*, 651–661. [CrossRef]
35. Xue, J.-h.; Li, W.; Qin, C. The scuffing load capacity of involute spur gear systems based on dynamic loads and transient thermal elastohydrodynamic lubrication. *Tribol. Int.* **2014**, *79*, 74–83. [CrossRef]
36. Im, K.; Avouac, J.-P. Linear stability analysis of the condition for vibration during frictional slip. *J. Mech. Phys. Solids* **2022**, *167*, 104993. [CrossRef]
37. Sui, X.; Ding, Q. Bifurcation and stability analyses for a pad-on-disc frictional system. *Int. J. Non-Linear Mech.* **2018**, *107*, 112–125. [CrossRef]
38. Onishchenko, V. Investigation of tooth wears from scuffing of heavy duty machine spur gears. *Mech. Mach. Theory* **2015**, *83*, 38–55. [CrossRef]
39. Parmar, A.; Ramkumar, P.; Shankar, K. Macro geometry multi-objective optimization of planetary gearbox considering scuffing constraint. *Mech. Mach. Theory* **2020**, *154*, 104045. [CrossRef]
40. Voitov, A.V. Parametric identification of the mathematical model of the functioning of tribosystems in the conditions of boundary lubrication. *Probl. Tribol.* **2021**, *27*, 6–14. [CrossRef]
41. Vojtov, V.; Biekirov, A.; Voitov, A. The quality of the tribosystem as a factor of wear resistance. *Int. J. Eng. Technol.* **2018**, *7*, 25–29. [CrossRef]
42. Voitov, A.; Fenenko, K.; Fenenko, O. Simulation of change in rheological properties of structure of combined materials in tribosystem. *IOP Conf. Ser. Mater. Sci. Eng.* **2021**, *1021*, 012052. [CrossRef]
43. Havrylenko, Y.; Kholodniak, Y.; Halko, S.; Vershkov, O.; Bondarenko, L.; Suprun, O.; Miroshnyk, O.; Shchur, T.; Śrutek, M.; Gackowska, M. Interpolation with Specified Error of a Point Series Belonging to a Monotone Curve. *Entropy* **2021**, *23*, 493. [CrossRef] [PubMed]
44. Iegorov, O.; Iegorova, O.; Miroshnyk, O.; Savchenko, O. Improving the accuracy of determining the parameters of induction motors in transient starting modes. *Energetika* **2020**, *66*, 15–23. [CrossRef]
45. Havrylenko, Y.; Kholodniak, Y.; Halko, S.; Vershkov, O.; Miroshnyk, O.; Suprun, O.; Dereza, O.; Shchur, T.; Śrutek, M. Representation of a Monotone Curve by a Contour with Regular Change in Curvature. *Entropy* **2021**, *23*, 923. [CrossRef]
46. Wei, Y.; Zhao, X.; Wei, Y.; Chen, Y.Q. Lyapunov stability criteria in terms of class K functions for Riemann–Liouville nabla fractional order systems. *ISA Trans.* **2022**, *131*, 137–145. [CrossRef]
47. Gokul, P.; Rakkiyappan, R. New finite-time stability for fractional-order time-varying time-delay linear systems: A Lyapunov approach. *J. Frankl. Inst.* **2022**, *359*, 7620–7631. [CrossRef]
48. Mondié, S.; Egorov, A.; Gomez, M.A. Lyapunov stability tests for linear time-delay systems. *Annu. Rev. Control.* **2022**, *54*, 68–80. [CrossRef]
49. Shen, C.; Li, Y.; Zhu, X.; Duan, W. Improved stability criteria for linear systems with two additive time-varying delays via a novel Lyapunov functional. *J. Comput. Appl. Math.* **2020**, *363*, 312–324. [CrossRef]
50. Zhan, X.; Hu, Y. On the relation between Hurwitz stability of matrix polynomials and matrix-valued Stieltjes functions. *J. Comput. Appl. Math.* **2023**, *417*, 114614. [CrossRef]
51. Zhan, X.; Dyachenko, A. On generalization of classical Hurwitz stability criteria for matrix polynomials. *J. Comput. Appl. Math.* **2021**, *383*, 113113. [CrossRef]
52. El-Marhomy, A.A.; Abdel-Sattar, N.E. Stability analysis of rotor-bearing systems via Routh-Hurwitz criterion. *Appl. Energy* **2004**, *77*, 287–308. [CrossRef]
53. Jin, X.-C.; Lu, J.-G. Delay-dependent criteria for robust stability and stabilization of fractional-order time-varying delay systems. *Eur. J. Control.* **2022**, *67*, 100704. [CrossRef]
54. Vojtov, V.; Fenenko, K.; Voitov, A.; Hrynkiv, A.; Lyashuk, O.; Vovk, Y. Methodical Approach to Using Acoustic Emission Method for Tribosystem Monitoring. *Tribol. Ind.* **2022**, *44*, 470–481. [CrossRef]

lubricants

MDPI

Article

Tribological, Oxidation and Thermal Analysis of Advanced Microwave—Hydrothermal Synthesised $Ti_3C_2T_x$ MXene as Additives in Outboard Engine Oil

Haizum Aimi Zaharin [1,*], Mariyam Jameelah Ghazali [1,*], Mohammad Khalid [2,3,4,5,*], Thachnatharen Nagarajan [6], Wong Weng Pin [2], Farah Ezzah [7], Ong Gerard [8], Rashmi Walvekar [9] and Abdul Khaliq Rasheed [10]

1 Department of Mechanical and Manufacturing Engineering, Faculty of Engineering and Built Environment, Universiti Kebangsaan Malaysia, Bangi 43600, Malaysia
2 Graphene and Advanced 2D Materials Research Group (GAMRG), School of Engineering and Technology, Sunway University, Petaling Jaya 47500, Malaysia
3 Sunway Materials Smart Science and Engineering (SMS2E) Research Cluster, Sunway University, Petaling Jaya 47500, Malaysia
4 School of Applied and Life Sciences, Uttaranchal University, Dehradun 248007, India
5 Division of Research and Development, Lovely Professional University, Phagwara 144411, India
6 Faculty of Defence Science and Technology, National Defence University of Malaysia, Kuala Lumpur 57000, Malaysia
7 Department of Chemical and Environmental Engineering, Malaysia-Japan International Institute of Technology (MJIIT), Universiti Teknologi Malaysia, Jalan Sultan Yahya Petra, Kuala Lumpur 54100, Malaysia; farahezzah.fea@gmail.com
8 Centre for Ionics University of Malaya, Department of Physics, Faculty of Science, Universiti Malaya, Kuala Lumpur 50603, Malaysia
9 Department Chemical Engineering, School of Energy and Chemical Engineering, Xiamen University Malaysia, Bandar Sunsuria, Sepang 43900, Malaysia; rashmi.walvekar@xmu.edu.my
10 Department of New Energy Science and Engineering, School of Energy and Chemical Engineering, Ximen University Malaysia (XMUM), Sepang 43900, Malaysia; abdulkhaliq.rasheed@xmu.edu.my
* Correspondence: p100279@siswa.ukm.edu.my (H.A.Z.); mariyam@ukm.edu.my (M.J.G.); khalids@sunway.edu.my (M.K.)

Abstract: In today's fast, globalised world, lubrication has become essential in enhancing engine efficiency, including in the marine sector. While the number of fishing vessels increased, so did the environmental pollution issues, due to inefficient engines. An outboard engine oil's tribological, oxidation and thermal conductivity behaviour play a crucial role in improving the quality of an outboard engine's life. In this research, $Ti_3C_2T_x$ MXene nanoparticles with different interlayer spacing were synthesised via an advanced microwave–hydrothermal approach. Later, the nanoparticles were dispersed in TC-W outboard engine oil to formulate the $Ti_3C_2T_x$ MXene nanolubricant with different concentrations. The results show that nanolubricant with a 0.01 wt.% $Ti_3C_2T_x$ MXene concentration with higher interlayer spacing reduced the coefficient of friction, and the average wear scar diameter by 14.5% and 6.3%, respectively, compared to the base oil. Furthermore, the nanolubricant with a 0.01 wt.% concentration of the $Ti_3C_2T_x$ MXene nanoparticle showed an improvement of 54.8% in oxidation induction time compared to the base oil. In addition, MXene nanolubricant established a more than 50% improvement in thermal conductivity compared to the base oil.

Keywords: MXene; tribology; nanolubricant; oxidation; thermal conductivity

1. Introduction

The rapidly expanding numbers of engine-powered fishing vessels, seafood industries and sea transport in international trade have resulted in a rise in the use of marine engines, which reward the world with beneficial effects on economic growth [1]. However, the offering does not come empty-handed, and there is a high price we have to pay. Human health, aquatic life, and the air and marine environment have all been compromised due to the poisonous emissions from inefficient marine engines and lubricants [1–4]. Therefore, improved marine engine oil is extremely important for enhancing engine performance and efficiency. Enhancing marine lubrication is expected to reduce wear and friction, extend the life of the mechanical components, increase fuel efficiency, and lower emissions. Furthermore, improved lubrication protects health and the environment, especially for populations living near ports and coasts, including aquatic species, in agreement with the United Nations Sustainable Development Goals (SDGs); in particular, for SDG 14, which is focused on "Life Below Water," it is crucial to explore innovative solutions that can mitigate marine pollution and promote sustainable practices, as well as SDG 13, climate action, SDG 14, underwater life and SDG 15, life on land [5].

Researchers have shifted their focus onto harnessing nanomaterials to enhance traditional lubricants and optimise engine performance, due to the rapid growth of nanotechnology. It is feasible to improve lubrication, boost fuel efficiency, and extend engine life by introducing nanotechnology nanoparticles into the outboard engine oil and ultimately contribute to reducing marine pollution [6,7]. Nanolubricants are specially engineered at the nanoscale, allowing them to offer enhanced lubricating properties compared to conventional lubricants. These nanoscale additives possess unique characteristics that minimise friction and wear between engine components [8–10]. By reducing friction, nanolubricants help optimise the engine's mechanical efficiency, improving fuel economy and reducing emissions. In addition to their superior lubricating properties, nanolubricants exhibit enhanced thermal stability. The nanoscale structure of these lubricants enables better heat dissipation, reducing the risk of thermal degradation and maintaining optimal operating temperatures within the engine [11,12]. This improved thermal stability translates to more efficient engine performance and longevity.

Nanotechnology is the manipulation and control of materials at the nanoscale level, which is generally between 1 and 100 nanometres. Materials have distinctive characteristics at this size that differ from their bulk counterparts. The capacity to manufacture nanoparticles with specific properties and functions has opened up new possibilities for improving lubricants in various applications [13]. The incorporation of nanoparticles into engine oil has various prospective benefits. Firstly, nanoparticles can minimise friction between moving engine parts by generating a protective coating that lowers metal-to-metal contact, thus reducing wear and tear [14,15]. Additionally, these nanoparticles can improve thermal conductivity, allowing for improved heat dissipation and lower operating temperatures. This characteristic is especially important for outboard engines used in harsh maritime conditions. Furthermore, the unique physicochemical features of nanoparticles can improve the engine oil's detergency and dispersibility, boosting its capacity to remove and suspend pollutants [16].

Furthermore, emerging research suggests that incorporating 2D MXene additives into marine engine oils holds great potential in addressing tribological problems [17] and significantly reducing pollution. MXenes are two-dimensional (2D) materials with layered transition metal carbides, nitrides, or carbonitrides. These 2D materials are typically prepared by etching initial materials known as MAX phase, represented by the generalised formula $M_{n+1}AX_n$. Here, "M" stands for the transition metals, "A" is an element from group 12 to 16 of the periodic table, "X" can be either carbon (C) or nitrogen (N), while "n" in this formula can range from 1 to 3 [18–21]. $Ti_3C_2T_x$ MXene, in particular, has received a lot of interest in recent years due to its exceptional features, including strong electrical conductivity, thermal stability, and excellent mechanical strength [22,23]. These properties make it an intriguing choice for adding to engine oil in order to improve lubrication and reduce engine friction. Dispersing $Ti_3C_2T_x$ MXene nanoparticles into engine oil has many potential advantages [24]. These nanoparticles can build a strong protective coating on the engine surfaces, minimising friction and wear between moving parts. The $Ti_3C_2T_x$ MXene's peculiar 2D structure helps it to adhere tightly to surface imperfections, encouraging improved lubrication and lowering the risk of surface damage.

Furthermore, MXenes' excellent thermal conductivity allows for effective heat dissipation, which aids in temperature management and prevents excessive engine overheating [25,26]. The $Ti_3C_2T_x$ MXene nanoparticles also have intrinsic tribological qualities, such as strong load-bearing capacity and good anti-wear capabilities. Their inclusion in engine oil can result in lower frictional losses, increased fuel economy, and a longer engine life [27].

Until recently, various synthesis approaches have been developed to achieve MXene production. Conventional wet-chemical etching at room temperature, involving hazardous HF etchant, offers a simple and accessible method but a toxic handling environment, and lacks control over morphology and homogeneity [28,29]. Hydrothermal and solvothermal heating methods enable the formation of crystalline MXenes at controlled temperatures and offer versatility, but they require specialised equipment and longer reaction times [30–32]. On the other hand, the electrochemical process provides precise control over MXene properties, but is limited to precursor materials stable under electrochemical conditions [33]. While the minimal intensive layer delamination (MILD) synthesis approach offers safer handling conditions and well-delaminated MXene sheets with preserved morphology, the formation of MXene sheets can be relatively slow, leading to extended synthesis durations [28,34]. Notably, the microwave-assisted hydrothermal method presents a compelling advantage in the characterisation of MXenes [34,35]. By utilising microwave irradiation during the hydrothermal synthesis, this method offers rapid synthesis times, improved crystallinity, enhanced homogeneity, and controlled morphology of the resulting MXenes [34,36–38]. According to Numan et al. [34], using microwaves together with the hydrothermal method facilitates the efficient energy transfer and heating of the reaction mixture, promoting the formation of high-quality MXene layers. Unlike the conventional hydrothermal method, the microwave-assisted hydrothermal technique combines heat from microwave irradiation and the hydrothermal process, providing a short heating time and reducing heat losses [39,40]. Furthermore, the rapid synthesis and low operating temperature of the microwave-assisted hydrothermal method, which is from 35 to 200 °C [35], hinders the exposure of MXene to long and high heating temperature, which possibly increases the tendency of oxidation, as in the hydrothermal approach.

This work adopted a novel method to synthesise $Ti_3C_2T_x$ MXene nanoparticles with a variable interlayer spacing, utilising an advanced microwave-hidrothermal synthesis platform which significantly decreased synthesis time and energy consumption. This study aims to examine and understand the underlying process by which the addition of $Ti_3C_2T_x$ MXene improves base oil performance. Various characteristics, such as the coefficient of friction (COF), average wear scar diameter (WSD), oxidation induction time (OIT), and thermal conductivity, were carefully investigated to attain this aim. The ultimate goal is to understand better the impacts of $Ti_3C_2T_x$ MXene additions on engine oil behaviour, which might lead to new high-performance lubricants. Using $Ti_3C_2T_x$ MXene additives, the findings of this study will provide essential insights into the performance enhancement of engine oil, which demonstrate great promise in mitigating tribological problems and overcoming pollution challenges. Through concerted efforts and interdisciplinary research, we can pave the way for a cleaner and more sustainable future for our marine ecosystems.

2. Materials and Methods

2.1. Materials

Materials used to prepare the MXene powder included titanium aluminium carbide (Ti_3AlC_2, Mesh 400, 99.9%) purchased from Xiamen Tob New Energy Technology Co., Ltd., Xiamen City, Fujian Province, China, hydrochloric acid, HCl (37% v/v, Merck, Rahway, NJ, USA), lithium fluoride, LiF (300 mesh), and ethanol (96%) purchased from Sigma Aldrich, Petaling Jaya, Selangor, Malaysia. The lubricant oil used was outboard engine oil with certified TC-W specifications by National Marine Manufacturers Association, NMMA. All chemicals were of analytical grade and were used without further purification.

2.2. Synthesis of $Ti_3C_2T_x$ MXene

$Ti_3C_2T_x$ MXene was synthesised using an advanced microwave hydrothermal synthesis platform (Milestone, flexiWAVE, Sorisole (BG), Italy). $Ti_3C_2T_x$ MXene was produced in situ HF by combining LiF and HCl. A total of 2.5 M LiF was added and dissolved in 10 mL of 6 M and 9 M concentrated HCl in a Teflon tube to prepare a 6 M-$Ti_3C_2T_x$ MXene and 9 M-$Ti_3C_2T_x$ MXene, respectively. A total of 0.5 g of MAX phase was progressively added to the LiF/HCl solution, to avoid risky exothermic reactions. The mixture was then magnetically stirred for 30 min, followed by 30 min of sonication. The combined solution was then transferred into an advanced microwave hydrothermal for time-effective etching. Using an advanced microwave-assisted hydrothermal approach, synthesis was carried out at 30 °C with a reaction time of 10 min. Figure 1 shows the illustration of $Ti_3C_2T_x$ MXene synthesis via advanced microwave–hydrothermal synthesis. When the reaction was completed, the acidic mixture was washed and rinsed a few times with deionised water and ethanol, before being centrifuged for 5 min at 5000 rpm using a Sartorius centrifuge (Goettingen, Germany). The supernatant was discarded, and the washing process was repeated until a stable black MXene colloidal solution was attained. The MXene solution, which is devoid of the aluminium layer and any subsequent products, was then taken as pure MXene. After that, this solution was applied to the MXene layer's delamination or opening process. In order to obtain the MXene powder, a freeze dryer was used to carry out the freeze-drying process. The synthesis methodology has been adopted from our previous work [34,41].

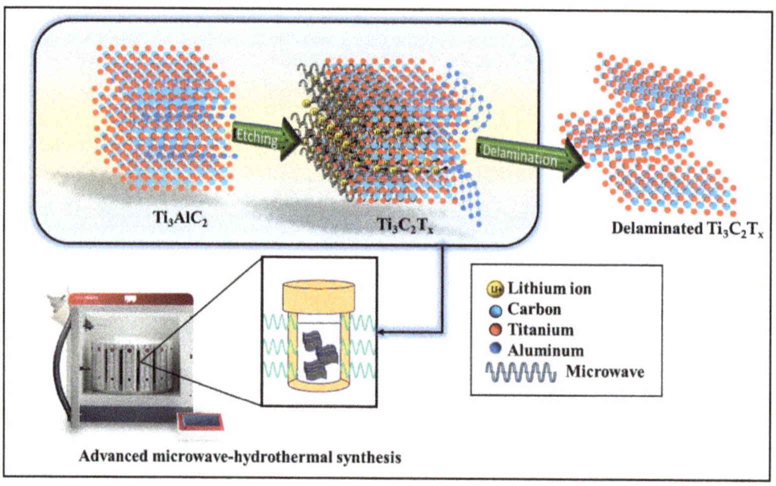

Figure 1. Schematic illustration of $Ti_3C_2T_x$ MXene synthesis via advanced microwave–hydrothermal synthesis.

2.3. Characterisation

The phase structure and crystallinity of as-synthesised 6 M-$Ti_3C_2T_x$ MXene and 9 M-$Ti_3C_2T_x$ MXene nanosheets were analysed using an X-ray diffractometer (XRD, BRUKER D8 advance) with Cu Kα radiation (U = 40 kV, I = 30 mA, and λ = 0.154 nm). The 2θ-degree patterns were observed between 5 and 80° at a 5°/min scanning rate. Energy-dispersive X-ray spectroscopy (EDX, Quanta 400F, Cambridge, MA, USA) was used to determine the elemental composition of the MXene. The morphology characterisation of $Ti_3C_2T_x$ MXene nanosheets was examined utilising field emission scanning microscopy (FESEM, ZEISS SUPRA 55VP).

2.4. Formulation of $Ti_3C_2T_x$ MXene Nanolubricant

The nanolubricant samples were formulated by dispersing 0.005 wt.%, 0.01 wt.%, and 0.05 wt.% as-synthesised 6 M-$Ti_3C_2T_x$ MXene and 9 M-$Ti_3C_2T_x$ MXene powder in 100 mL TC-W certified outboard engine oil by NMMA. In order to ensure homogenised dispersion of nanoparticles and obtain stability, oil samples were subjected to sonication using a water bath sonicator for 30 min and further homogenised for 10 min utilising a high shear lab mixer. For each concentration, three samples were prepared and repeatedly measured to minimise the errors and inconsistencies, ensure reliability and maintain quality control.

2.5. Physiochemical Characterisation of $Ti_3C_2T_x$ MXene Nanolubricant

A stability test was conducted using Zeta Potential (Malvern Zetasizer 3000HSA) for each oil sample, to ensure the stability of $Ti_3C_2T_x$ MXene nanolubricant and no MXene flocculation build-up in the oil. A viscometer (Viscometer SWM 3000, Anton Paar, Graz, Austria) was utilised to measure the density, kinematic viscosity, and viscosity index of the formulated 6 M-$Ti_3C_2T_x$ MXene and 9 M-$Ti_3C_2T_x$ MXene nanolubricant. The kinematic viscosity of the oil samples was measured at temperatures of 40 °C and 100 °C. Kinematic viscosity in the present study was calculated according to Equation (1):

$$v = \frac{\mu}{\rho} \tag{1}$$

where ν, μ and ρ represent kinematic viscosity, dynamic viscosity and density, respectively. The viscosity index (*VI*) is the rate of viscosity change as a function of temperature. In essence, *VI* is necessary to determine whether the lubricant satisfies the asset's requirements, depending on the operating temperature range. The *VI* was determined according to the standard ASTM D2270 and calculated using Equation (2):

$$VI = \frac{L - U}{L - H} \times 100 \tag{2}$$

where U denotes the oil's kinematic viscosity at 40 °C, whereas L and H are the reference oil's kinematic viscosities at 40 °C and 100 °C, respectively, as determined by ASTM D2270.

2.6. Four-Ball Tribotesting of $Ti_3C_2T_x$ MXene Nanolubricant

The tribological testing was conducted using a four-ball tribometer (DUCOM), following standard testing ASTM D 4172, to investigate the friction and wear properties of nanolubricant with different $Ti_3C_2T_x$ MXene concentrations. The temperature used was 75 °C, the rotational speed was 1200 rpm, and the applied load given to the other three balls was 392.5 N. The test was performed for 3600 s. Carbon–chromium steel balls were used in this test. Table 1 shows the mechanical properties of the balls used. After the testing, the image of the wear scar on the metal balls was analysed using field emission scanning electron microscopy and energy-dispersive X-ray spectroscopy (FESEM and EDX, Quanta 400F, Cambridge, MA, USA). The mechanical properties of the carbon-chromium steel balls used in the test are shown in Table 1.

The coefficient of friction was calculated using Equation (3):

$$\mu = 2.22707\frac{\tau}{\rho} \tag{3}$$

where μ represents the coefficient of friction, τ denotes the average frictional torque in kg-cm, and ρ is the load applied while performing the tribotest.

Table 1. Mechanical properties of the carbon-chromium steel ball-bearing used in four-ball tribotesting.

Mechanical Properties	Value
Hardness (H)	1 HRC
Density (ρ)	7.79 gm/cm^3
Surface roughness (Ra)	0.22

2.7. Oxidation Analysis of $Ti_3C_2T_x$ MXene Nanolubricant

MXene nanolubricants and outboard engine oil samples were measured to determine their oxidation induction time (OIT). The differential scanning calorimetry (DSC) method, following standard procedure ASTM 6186, was utilised to study the oxidative stability of nano marine oil. The tests were conducted using high-pressure differential scanning calorimeters (HP-DSC 250, from TA Instrument, Cheshire, UK). In this experiment, oxygen pressure was set and maintained at 200 psi throughout the experiment. A total of 3 mg of each sample was inserted into a standard aluminium pan as a test cell, while an empty pan was used as a reference. Both were used for oxidation testing and placed in a DSC cell. The DSC lid was closed, and oxygen was purged into the system until the pressure reached 200 psi. In the isothermal procedure, the measurements were performed at a temperature of 185 °C for 70 min. The DSC curve appeared in the form of exothermic heat flow during the initial state of the oxidation reaction, indicating the oxidation process. The oxidation induction time, which is the time required to begin the oxidation process of a sample, was determined from the DSC curve. Based on this curve, the point of intersection of the extrapolated baseline and the tangent line, which is the leading edge of the exothermal peak, represented the OIT.

2.8. Thermal Conductivity Analysis of $Ti_3C_2T_x$ MXene Nanolubricant

The good thermal conductivity of lubricating oil is the key to the excellent performance of the lubricating oil. The thermal conductivity of lubricating oil samples with different concentrations of $Ti_3C_2T_x$ MXene was evaluated using laser flash analysis, LFA HyperFlash, NETZSCH, Germany. In this experiment, samples were filled into the sample ring evenly. Before the sample holder components were assembled, the top and bottom sealing discs were sprayed with graphite to promote absorption. The sample was then subjected to laser flash analysis, LFA. Heating was applied from room temperature to 100 °C at a rate of 10 °C/min. The experiment was conducted in a nitrogen atmosphere.

3. Results and Discussions

3.1. Chemical and Structural Characterization of Nanomaterials

The phase structure, crystallinity, morphology and elemental composition of bulk MAX phase precursor and as-synthesised 9 M-$Ti_3C_2T_x$ MXene, as well as 6 M-$Ti_3C_2T_x$ MXene nanosheets, were characterised by XRD, FESEM, and EDX; these are presented in Figure 2. The XRD patterns in Figure 2A were observed to validate the formation of 9 M-$Ti_3C_2T_x$ MXene and 6 M-$Ti_3C_2T_x$ MXene from MAX phase Ti_3AlC_2. The XRD result evidently exhibits that the $Ti_3C_2T_x$ MXene is effectively attained from the etching and exfoliating of the MAX phase through advanced microwave-assisted hydrothermal synthesis. The XRD data for the MAX phase diffraction (002), (004), (101), (103), (104), (105), (109), and (110) planes are observed at 9.7°, 19.2°, 34.1°, 39.0°, 41.9°, 56.5°, and 60.4°, respectively, consistent with the standard (JCPDS No.52-0875). After going through the advanced microwave–hydrothermal synthesis, most of the sharp peaks at Ti_3AlC_2 almost disappeared, and a sharp diffraction peak at 9.64° 2θ, corresponding to the (002) diffraction plane, are gradually shifted from 9.64° to 6.13° and 6.54° for 9 M-$Ti_3C_2T_x$ MXene and 6 M-$Ti_3C_2T_x$ MXene, respectively. The (002) plane broadened and shifted towards a lower angle, suggesting that the grain size reduced and the nanosheet spacing of MXene increased, in agreement with the previous literature [34,42]. According to the Bragg formula, the c-lattice parameters for 9 M-$Ti_3C_2T_x$ MXene and 6 M-$Ti_3C_2T_x$ MXene are about 24.73 Å and 22.10 Å, respectively. The result indicates that, with higher HCl concentration, in this case 9 M, more H^+ ions are provided by HCl to react with fluoride salt to form HF. Sufficient HF eases the etching of Al and expands the interlayer spacing between MXene sheets easily [43]. Furthermore, a higher amount of H^+ in 9 M HCl promotes hydrophilicity. H^+ attracts more water molecules to intercalate between MXene nanolayers during etching, resulting in an increased expansion of interlayer spacing compared to the lower concentration of HCl. Figure 2B,C,F depict the morphological structure of MAX phase Ti_3AlC_2, 9 M-$Ti_3C_2T_x$ MXene and 6 M-$Ti_3C_2T_x$ MXene, respectively, after going through etching and exfoliation with HCl/LiF via advanced microwave-assisted hydrothermal synthesis. In Figure 2B, the bulk MAX phase precursor Ti_3AlC_2 microstructure shows a solid and stacked multilayer nanosheet structure. In contrast, in Figure 2C,F, crumpled MXene sheets can be seen, signifying the exclusion of the Al element from packed MAX layers throughout the etching and exfoliation. The result demonstrates an effective formation of thin layers of 2D MXene nanosheets using an advanced microwave-assisted hydrothermal approach. The elemental mapping of both MXenes in Figure 2D,G shows the uniform spatial distribution of Ti, C, F, Cl, O, and Al. To validate the XRD and FESEM result, the EDX elemental spectrum in Figure 2E,H demonstrates that both MXene samples contain Ti, C, F and a small amount of O, Cl, and Al elements. The EDX analysis verifies the formation of MXene and the removal of most Al elements from the MAX phase.

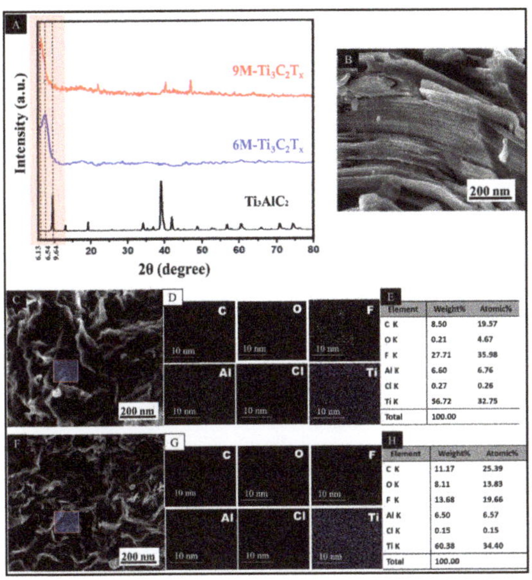

Figure 2. Structural characterisation of as-synthesised $Ti_3C_2T_x$ MXene. (**A**) XRD diffractogram of MAX phase, 9 M-$Ti_3C_2T_x$ MXene and 6 M-$Ti_3C_2T_x$ MXene. (**B**) FESEM micrographs of MAX phase. (**C**) FESEM micrographs of 9 M-$Ti_3C_2T_x$ MXene. (**D**) 9 M-$Ti_3C_2T_x$ MXene corresponding EDX elemental mapping. (**E**) Elemental composition of 9 M-$Ti_3C_2T_x$ MXene calculated from EDX analysis. (**F**) FESEM micrographs of 6 M-$Ti_3C_2T_x$ MXene. (**G**) Corresponding EDX elemental mapping of 6 M-$Ti_3C_2T_x$ MXene. (**H**) EDX spectra of 9 M-$Ti_3C_2T_x$ MXene and 6 M-$Ti_3C_2T_x$ MXene, respectively.

3.2. Physiochemical Characterisation of Nanolubricant

The zeta potential measures the strength of the electrokinetic potential, such as the attractive and repulsive interactions between particles suspended in a dispersion [44,45]. The size of the zeta potential value will determine whether the dispersion is relatively stable. The larger the absolute magnitude of the zeta potential, the stronger the dispersion will resist aggregation, resulting in a more extended period of stability. On the other hand, the dispersion is more likely to coagulate, and its stability time is shortened when the absolute value of the zeta potential is closer to 0 mV. Figure 3A explains the stability ranges of the zeta potential. For zeta potential values between 0 and ±5 mV, the dispersed phase strongly tends to agglomerate; for values between ±10 and ±30 mV, an incipient instability is indicated; for values between ±30 and ±40 mV, moderate stability is identified; for values between ±40 and ±60 mV, good stability is specified; and for values higher than ±61 mV, excellent stability is signified. Thus, nanolubricant stability is significantly affected by their electrokinetic properties. Due to the strong repulsive forces exerted by the high surface charge density, the probability of flocculation is reduced [45–47].

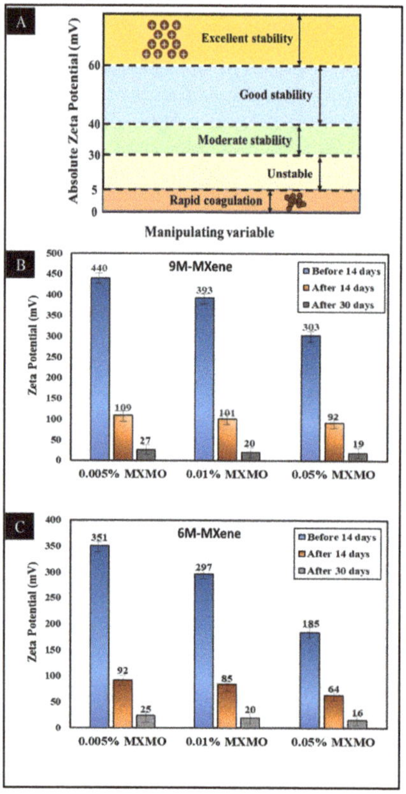

Figure 3. (**A**) Schematic illustration explaining stability and zeta potential ranges. (**B**,**C**) The zeta potential of 9 M-$Ti_3C_2T_x$ MXene and 6 M-$Ti_3C_2T_x$ MXene nanolubricants before 14 days, after 14 days, and after 30 days.

Figure 3B shows the zeta potential of 9 M-$Ti_3C_2T_x$ MXene and 6 M-$Ti_3C_2T_x$ MXene nanolubricant before 14 days, after 14 days, and after 30 days. In this study, the zeta potential of MXene nanolubricant dispersion shows >61 mV, presenting the excellent stability of both 9 M-$Ti_3C_2T_x$ MXene and 6 M-$Ti_3C_2T_x$ MXene, regardless of the concentration, before and after 14 days. However, the stability of MXene nanolubricant deteriorates after 30 days, where zeta potential is less than 30 mV, indicating coagulation of MXene nanoparticles occurred after one month. It is also observed that the zeta potential value almost linearly decreased with $Ti_3C_2T_x$ MXene concentration. This suggests that there is restricted nanoparticle mobility at higher concentrations, which inhibits the building of the energy barrier and causes particle agglomeration and sedimentation. In other words, the stearic repulsion between nanoparticles rises with concentration, causing them to aggregate and reduce the zeta potential value. According to Said et al. [44], when particle concentration rises, the average static surface-to-surface spacing decreases, which causes the particle to cluster. Thus, high spacing between the particles for 9 M-$Ti_3C_2T_x$ MXene and 6 M-$Ti_3C_2T_x$ MXene would be ineffective.

Figure 4 presents the kinematic viscosity of MO, 9 M-$Ti_3C_2T_x$ MXene, and 6 M-$Ti_3C_2T_x$ MXene nanolubricants tested at 40 °C and 100 °C. The results show a trend of reduction in kinematic viscosity when the temperature rises. With the increment in temperature, the thermal energy induces higher kinetic energy towards molecules, which causes rapid and mobile movement. The attractive binding energy between the molecules decreases, significantly reducing the kinematic viscosity [11,48]. Therefore, the kinematic viscosity findings show that MXene nanoparticles assist in modifying the viscosity of the nanolubricant when the temperature changes. In addition, the kinematic viscosity shows little increase after adding different concentrations of MXene nanoparticles. There are no significant changes between the kinematic viscosity of 9 M-$Ti_3C_2T_x$ MXene and 6 M-$Ti_3C_2T_x$ MXene nanolubricants, due to minimal differences in interlayer spacing of MXene nanosheets, which is considered a negligible factor influencing the kinematic viscosity of the lubricant. Thus, the reduction in kinematic viscosity suggests that MXene nanoparticles work as an increasing catalyst as the concentration of MXene increases, regardless of the size of the c-lattice parameter.

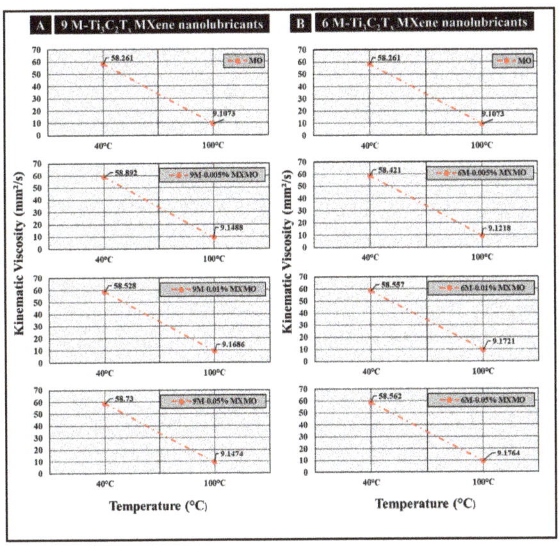

Figure 4. Kinematic viscosity at 40 °C and 100 °C of (**A**) 9 M-$Ti_3C_2T_x$ MXene nanolubricant and (**B**) 6 M-$Ti_3C_2T_x$ MXene nanolubricant.

Furthermore, the demand for engine oils with higher fuel-saving efficiency has been driven by the need to reduce CO_2 emissions from marine transportation. The implementation can be anticipated by developing an engine oil with a high viscosity index (VI). In general, engine oils become more viscous at lower temperatures, increasing engine drag. High VI reduced oil viscosity at lower temperatures, improving fuel-saving efficiency. Figure 5 exhibits the variation in viscosity index with different concentrations of 9 M-$Ti_3C_2T_x$ MXene and 6 M-$Ti_3C_2T_x$ MXene nanolubricants. Figure 5 reveals that adding 0.01 wt.% of 9 M-$Ti_3C_2T_x$ MXene and 6 M-$Ti_3C_2T_x$ MXene elevated the VI of base oil by 1.14% and 0.76%, respectively. Compared to base oil, MXene nanolubricants show slightly higher VI, which offers low viscosity change at varying temperatures, improved thinning resistance, and preserved protective oil film strength under high pressure and temperature. This finding demonstrates that using MXene nanoparticles in outboard engine oil potentially enhances the viscosity index and increases engine oil preference for high-temperature use.

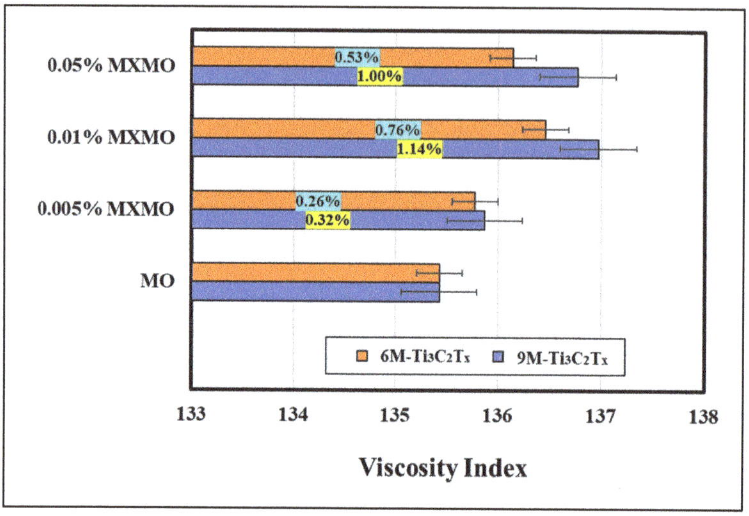

Figure 5. Viscosity index of the nanolubricant with 9 M-$Ti_3C_2T_x$ MXene and 6 M-$Ti_3C_2T_x$ MXene nanolubricant.

3.3. Tribological Analysis of $Ti_3C_2T_x$ MXene Nanolubricant

As is well known, adding excessive or insufficient nano-additives with the ideal interlayer spacing of MXene can cause lubricating oil performance to be compromised. The coefficient of friction (COF) and average wear scar diameter (WSD) of MXene nanolubricant with different interlayer spacing and concentrations of $Ti_3C_2T_x$ MXene nanoparticles in outboard engine oil are therefore used to determine the optimal interlayer spacing and concentration. The results are shown in Figure 6. As depicted in Figure 6A, the COF of the base oil is 0.1252. The COF first decreased when 9 M-0.005 wt.% and 9 M-0.01 wt.% of MXene were added to the base oil. Nanolubricant with 9 M-0.005 wt.% of $Ti_3C_2T_x$ MXene contains an adequate number of nanoparticles to promote formation of a protective layer between the mating surfaces. Building a tribofilm comprised of nanosheets helps mitigate the friction caused by the slippage of the individual layers of nanosheets [49]. Further addition of MXene concentration, 9 M-0.01 wt.%, shows a significant decrement of COF, possibly attributed to the optimal amount of nanosheets to form a homogenous tribofilm during sliding and deformation of individual nanosheets, which results in a reduction in friction between mating surfaces. However, the COF started to increase when the concentration reached up to 9 M-0.05 wt.%, which can be attributed to the flocculation of $Ti_3C_2T_x$ MXene nanoparticles, dramatically expanding the nanoparticle size. Consequently, the nanoparticles cannot enter the tiny space between the friction contact, leading to higher COF [8,49,50].

It should be noted that adding 6 M-$Ti_3C_2T_x$ MXene, with lower interlayer spacing, into base oil exhibits the same COF reduction trend, with an insignificant percentage of COF decrement (less than 10%). This may be due to deficient interlayer spacing to facilitate lower shear strength, which promotes better lubricity to reduce friction, compared to 9 M-$Ti_3C_2T_x$ MXene nanoparticles with higher interlayer spacing. On the other hand, 6 M-0.01 wt.% $Ti_3C_2T_x$ MXene nanolubricant also demonstrates the highest improvement in COF amongst other 6 M concentrations, suggesting that 0.01 wt.% of MXene is the optimal amount to establish a uniform tribolayer between mating surfaces. Overall, MXene additive nanolubricants with varying concentrations and interlayer spacing assist in reducing COF and wear, showing potential as an anti-friction and anti-wear agent. In evaluation with other reported studies, a comparison table showing our results and the tribological performance of other nanomaterials is attached in Appendix A.

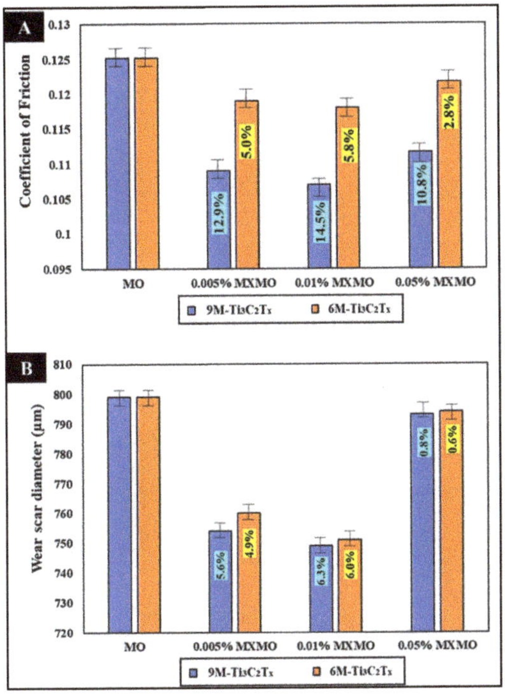

Figure 6. (**A**) The coefficient of friction of 9 M-$Ti_3C_2T_x$ MXene and 6 M-$Ti_3C_2T_x$ MXene nanolubricants. (**B**) The average wear scar diameter of 9 M-$Ti_3C_2T_x$ MXene and 6 M-$Ti_3C_2T_x$ MXene nanolubricants.

Figure 6B presents the average wear scar diameter (WSD) with varying MXene additive concentrations in outboard engine oil. The WSD of the ball without any addition of MXene additive is 818 µm. Figure 6B shows that 0.005 wt.%–0.05 wt.% of 9 M-and 6 M-$Ti_3C_2T_x$ MXene additive dispersed in outboard engine oil improved WSD from 0.6% to 6.3% in comparison with base oil, while higher improvement in WSD is shown by 9 M-$Ti_3C_2T_x$ MXene nanolubricant compared to 6 M-$Ti_3C_2T_x$ MXene nanolubricant; however, no significant difference in WSD reduction associated to the variation in interlayer spacing. The result also demonstrates 0.01 wt.% 9 M-$Ti_3C_2T_x$ MXene additive establishes the lowest average WSD. The decrease of WSD of 0.01 wt.% 9 M-$Ti_3C_2T_x$ MXene is attributed to the forming of a thin lubricating layer between the contact surfaces, which helps minimise contact pressure and frictional torque [47]. Furthermore, the occurrence of the mending effect can promote the reduction in the average WSD. This effect occurs when the 2D nanoparticles accumulate and deposit in the tiny cracks and ridges of mating surfaces, smoothing them out and lowering the average WSD. This mechanism and result are in agreement with former studies [8,41,51,52].

These findings were further confirmed with FESEM and EDX analyses on the steel ball surface, as shown in Figure 7. In Figure 7A, the surface of the ball bearing with base oil with no addition of MXene shows darker concentric grooves, indicating a deeper furrow. The lowest WSD was chosen to compare with the WSD of a non-additive nanolubricant. In Figure 7B, with 0.01 wt.% 9 M-$Ti_3C_2T_x$ MXene additive, the scar under FESEM observation appears to be brighter, representing a shallower furrow. Bright and smooth wear tracks appeared when the lubricant with $Ti_3C_2T_x$ MXene nanoparticles was mixed, demonstrating a reduction in the contact surfaces between the steel balls. It is hypothesised that MXene nanosheets can easily slide and infiltrate the oil surface due to their two-dimensional structure. Additionally, nanosheets offer a continuous layer on sliding surfaces because of their excellent contact adherence. This behaviour is associated with the mending property of $Ti_3C_2T_x$ MXene. The nanoparticles repair the scratched and worn surfaces by being deposited into worn surfaces, while building a protective layer to eliminate direct contact between the two contact surfaces, minimising the wear scar diameter. The adherence of $Ti_3C_2T_x$ MXene nanoparticles on the cracks assists in a reduction in the depth of the grooves, as evidenced in Figure 7D by the presence of Ti element in the EDX elemental spectrum of the scar on the steel ball. In comparison, no Ti element was observed on the steel ball scar with the base oil lubricant in Figure 7C. The experimental tribological results above suggest that, with an optimal concentration of 0.01 wt% $Ti_3C_2T_x$ MXene in the outboard engine oil, COF and WSD can be significantly improved.

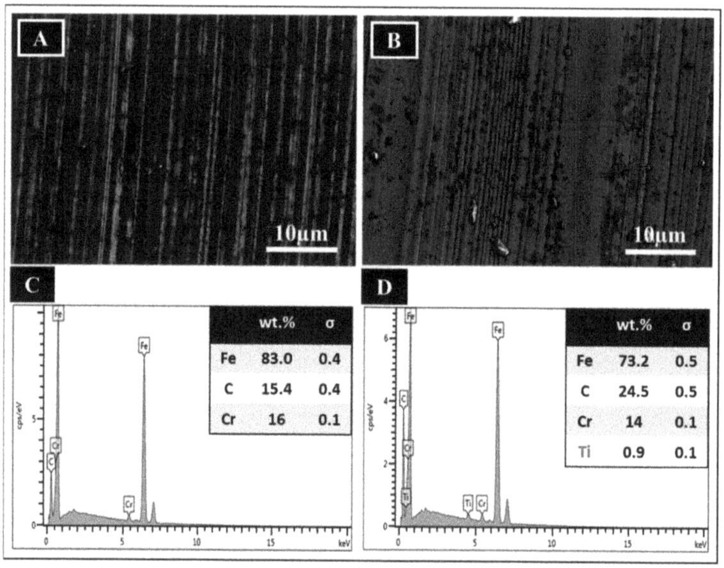

Figure 7. FESEM images of wear scar diameter with (**A**) base oil and (**B**) 9 M-0.01 wt.% MXMO. EDX elemental spectra of the wear scar with (**C**) base oil and (**D**) 9 M-0.01 wt.% MXMO.

According to the previous study, the development of tribofilm and the mending effect is the fundamental mechanism for reducing frictional wear in the case of $Ti_3C_2T_x$ MXene nanolubricant. The two-dimensional plane structure of $Ti_3C_2T_x$ MXene allows it to glide between the oil surface easily. Furthermore, when the concentration of $Ti_3C_2T_x$ MXene grows, it will agglomerate and precipitate, increasing wear and friction between mating surfaces. The wear process of $Ti_3C_2T_x$ MXene nanosheets is related to the segregation of interlayers into different layers [53,54] due to reduced van der Waals or Coulombic repulsive interactions at contact points. Thus, a higher interlayer spacing with lower van der Waals interaction easily disintegrated into individual layers and quickly adhered to the worn surface, promoting better lubricity to the contact surface, resulting in higher improvement in COF and wear compared to $Ti_3C_2T_x$ MXene with lower interlayer spacing, as supported by the previous literature [55]. Thus, the findings show that adding $Ti_3C_2T_x$ MXene to the lubricant substantially enhances its tribological properties.

3.4. Oxidation Analysis of $Ti_3C_2T_x$ MXene Nanolubricant

High loads and temperatures and continuous air contact are the main contributors to the oxidation of lubricants in the transportation industries, including marine transportation. Oxidation accelerates the degradation of base oils and additives, decreasing their efficiency, performance, and life expectancy. Figure 8 displays the oxidation induction time (OIT) of 9 M-$Ti_3C_2T_x$ MXene nanolubricant since 9 M-$Ti_3C_2T_x$ MXene nanolubricant consistently shows better physiochemical and tribological properties compared to 6 M-$Ti_3C_2T_x$ MXene nanolubricant. The results show OIT enhancement by adding 9 M-$Ti_3C_2T_x$ MXene to the outboard engine oil. Compared with base oil, the OIT improves by 35.1%, 54.8% and 30.1% for 0.005 wt.%, 0.01 wt.% and 0.05 wt.%, respectively. Amongst other concentrations of nanolubricant formulation, the nanolubricant with 0.01 wt.% 9 M-$Ti_3C_2T_x$ MXene nanoparticles was found to enhance theOIT by the most suggesting that this concentration is the optimal concentration of MXene additive in the outboard engine oil, which offers the anti-oxidation effect that can expand lubricant service lifespan.

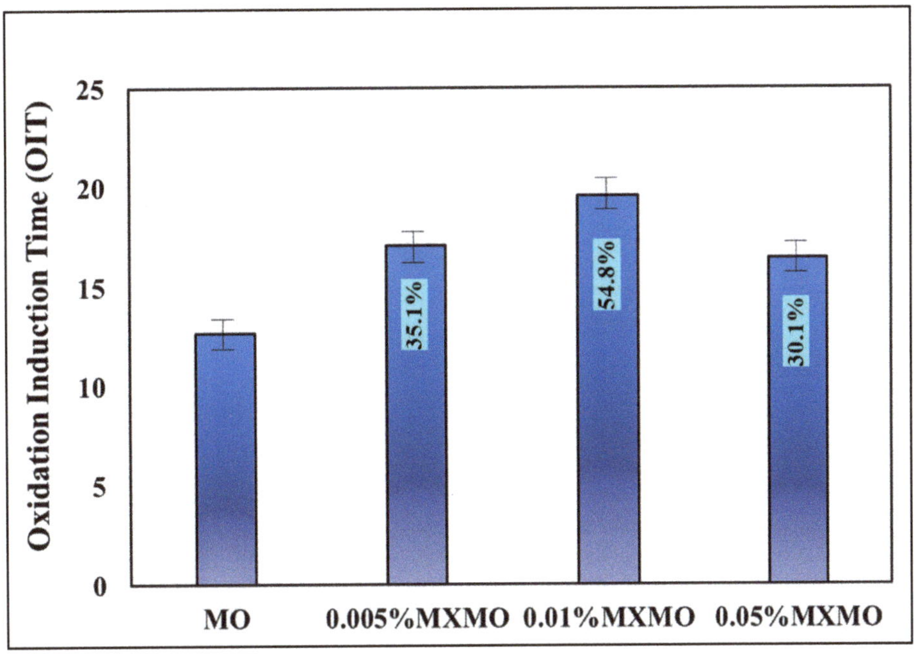

Figure 8. The oxidation induction time (OIT) of nanolubricants with different MXene concentrations.

3.5. Thermal Conductivity Analysis of $Ti_3C_2T_x$ MXene Nanolubricant

In order to reduce the heat generated by the moving components of an engine and overcome mechanical fatigue, a good lubricant is required to possess a high level of thermal conductivity [12]. The thermal conductivities of $Ti_3C_2T_x$ MXene with TC-W certified outboard engine oil samples for different concentrations are plotted in Figure 9A as a function of temperature from 40 °C to 100 °C. The influence of concentration and temperature on the thermal conductivity of $Ti_3C_2T_x$ MXene nanolubricant was analysed. The standard deviations for the thermal conductivity of the samples are ±0.002. The result shows that the thermal conductivity of $Ti_3C_2T_x$ MXene nanolubricant samples almost linearly increased with $Ti_3C_2T_x$ MXene concentration from 0.005 wt.% to 0.05 wt.% in engine oil. The high concentration of $Ti_3C_2T_x$ MXene dramatically increases with the content of $Ti_3C_2T_x$ MXene nanoparticles in the oil sample. The sample of oil with 0.05 wt.% of $Ti_3C_2T_x$ MXene exhibits the maximum thermal conductivity, which is 0.14 W/m K at 40 °C, and increases to 0.28 W/m K at 100 °C, with 59% and 69.7% of thermal conductivity enhancement, respectively, as shown in Figure 9B. With higher concentrations of MXene and uniformly dispersed ultra-high thermal conductivity nanoparticles, the nanoparticles are easily connected with each other to enhance heat dissipation. In other words, the molecular collisions between the base oil and more nanoparticles led to higher thermal conductivity. [49,56,57]. In addition, $Ti_3C_2T_x$ MXene nanosheets offer a high aspect ratio due to the size and layer morphology, which provide an increased contact interface to improve the thermal conductivity of lubricants with a higher amount of MXene nanoparticles [12,48]. Furthermore, the improvement in thermal conductivity of $Ti_3C_2T_x$ MXene nanolubricant is associated with the vast basal plane of $Ti_3C_2T_x$ MXene sheets present in outboard engine and the intense heat conductivity across the basal plane of $Ti_3C_2T_x$ MXene nanoflakes [58,59]. These planes have a high density of delocalised electrons, and can transfer heat quickly and efficiently through the material, which significantly improves the thermal conductivity of $Ti_3C_2T_x$ MXene nanolubricant.

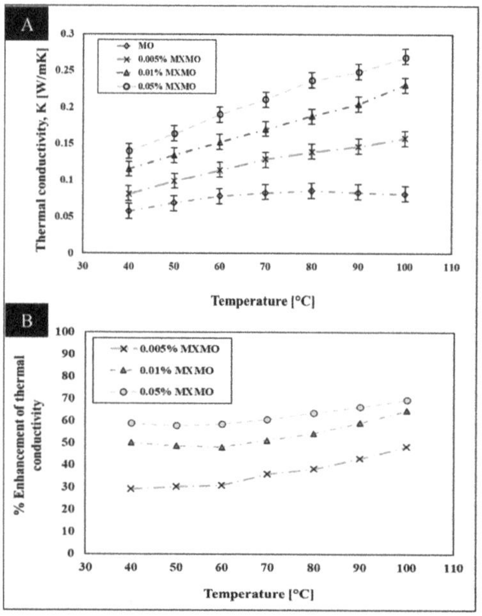

Figure 9. (**A**) Thermal conductivity of $Ti_3C_2T_x$ MXene with outboard engine oil as a function of temperature for the varying MXene concentrations of 0.005, 0.01, and 0.05 wt.%. (**B**) Percentage of thermal conductivity enhancements of $Ti_3C_2T_x$ MXene with outboard engine oil as a function of temperature for different concentrations.

Figure 9 shows that thermal conductivity also increases with temperature elevation. The molecules in base oil obtain a high-velocity vibration as the temperature rises. Dispersing $Ti_3C_2T_x$ MXene nanoparticles in outboard engine oil also induces a random collision; striking and hitting nanoparticle surfaces causes the Brownian motion effect. Under the high-temperature condition, the $Ti_3C_2T_x$ MXene nanoparticles gain more energy, and the Brownian motion effect becomes more intense, which results in a fascinating phenomenon known as thermophoresis [49,59,60]. Thermophoresis is the movement of particles brought on by a temperature gradient in the fluid. Brownian motion and thermophoresis both promote particle collision, though thermophoresis may sometimes have a more substantial impact. These particle collisions often result in an increase in thermal conductivity.

It is also worth noting that, at low temperatures, the lubricant's viscosity is higher, which can limit the mobility of the $Ti_3C_2T_x$ MXene nanosheets in the lubricant. This can decrease the effect of Brownian motion and the lubricant's thermal conductivity because the $Ti_3C_2T_x$ MXene nanoparticle transport between the MXene layers is decelerated. As the temperature increases, the lubricant's viscosity decreases, allowing for greater mobility of the $Ti_3C_2T_x$ MXene nanosheets. This leads to an increase in the thermal conductivity of the lubricant attributed to the high Brownian motion effect [61–63].

4. Conclusions

The current work has effectively established $Ti_3C_2T_x$ MXene nanoparticles as a potent anti-friction, anti-oxidant, and excellent thermal conductor additive with improved tribological, oxidation, and thermal conductivity performance in outboard engine oil. In the tribological analysis, nanolubricant with 0.01 wt.% concentration of $Ti_3C_2T_x$ MXene nanoparticles shows outstanding results in reducing the friction coefficient and average wear scar diameter, with 14.5% and 6.3% decrement, respectively, compared to the base oil. This is due to the tribofilm and mending effect, which promotes the formation of a protective film between the frictional surfaces and the deposition of nanoparticles in the cracks and ridges, decreasing the COF and average WSD. Furthermore, the addition of 0.01 wt.% $Ti_3C_2T_x$ MXene nanoparticles depict significant increases in OIT by 54.8% compared to base oil, revealing its anti-oxidation effect that can extend lubricant service duration. For the thermal conductivity analysis, the deployment of $Ti_3C_2T_x$ MXene nanoparticles shows an increasing trend in thermal conductivity as the number of nanoparticles increases. This can be explained by the high aspect ratio and the large basal plane of $Ti_3C_2T_x$ MXene nanoparticles, the effect of Brownian motion, and the thermophoresis phenomenon, which enhanced the thermal conductivity with increasing nanoparticles and temperature.

Author Contributions: Conceptualisation, H.A.Z., M.J.G., M.K. and A.K.R.; methodology, H.A.Z., M.K., T.N. and A.K.R.; formal analysis, H.A.Z., W.W.P. and R.W.; writing—original draft preparation, H.A.Z.; writing—review and editing, T.N., F.E. and O.G.; visualisation, H.A.Z., F.E. and O.G.; supervision, M.J.G., M.K. and A.K.R.; project administration, M.J.G. and M.K.; funding acquisition, M.J.G. and M.K. All authors have read and agreed to the published version of the manuscript.

Funding: This research was funded by the Ministry of Higher Education Malaysia and Universiti Kebangsaan Malaysia (grant number FRGS/1/2018/TK03/UKM/02/8) and also the international network research grant scheme from Sunway University, Malaysia (STR-IRNGS-SET-GAMRG-01-2022).

Data Availability Statement: Not applicable.

Conflicts of Interest: The authors declare no conflict of interest.

Appendix A

Material	Base Fluid	Concentration	Enhancement of Tribological Properties		Reference
			COF	Wear	
MXene $Ti_3C_2T_X$ nanosheets	DI- H_2O (AP30-$Ti_3C_2T_X$)	0.8 mg/mL	34.74%	45.58%	[64]
MXene $Ti_3C_2T_X$	DI- H_2O	5 wt.%	20%	48%	[65]
MXene $Ti_3C_2T_X$	PAO8 base oil	0.8 wt.%	9.5%	7.7%	[66]
$Ti_3C_2T_X$	Outboard oil	0.01 wt.%	0.8%	-	[1]
$TiO_2/Ti_3C_2T_X$	Oil	1.0 wt.%	29.1% (with 20N load)	Depth: 30 μm Width: 280 μm	[67]
$Ti_3C_2(OH)_2$	100SN base oil	1.0 wt.%	around 20% (with 15N load)	-	[68]
$Ti_3C_2T_X$	Liquid paraffin	1.0 wt.%	49.6%	-	[69]
KTO-$Ti_3C_2T_X$	PAO8 base oil	1.0 wt.%	30.6%	-	[55]

TDPA-Ti_3C_2	Castor oil	0.1 wt.%	27.9%	55.1%	[70]
DDP-Ti_3C_2Tx	500SN base oil	0.3 wt.%	COF: 0.11	87%	[71]
MXenes/MoS_2 heterojunction	Liquid paraffin	5.0 wt.%	39%	-	[72]
$Ti_3C_2T_X$/MoS_2 heterojunction	150SN base oil	0.3 wt.%	39%	85%	[73]
MXene-HS	PAO 10	1.0 wt.%	COF: 0.12	82%	[74]
Hybrid MoS_2@Ti_3C_2	SAE 5W-40-based engine oil	0.05 wt.%	13.9%	23.8%	[47]
MoS_2	SAE 20W50 diesel engine oil	0.01 wt.%	19.24%	19.52%	[49]
WS_2	Base oil + PVP surfactant	1.0 wt.%	33%	45%	[75]
Multi-layered graphene	PAO2 oil	0.05 wt.%	78%	16%	[76]
Graphene nanoplatelets (GNPs)	Palm oil TMP	0.05 wt.%	5%	15%	[77]
Graphene nanoparticle (GP)	Synthetic oil, PAO4	0.01 wt.%	78% (at 60–100 °C)	90% (at 60–100 °C)	[78]
Graphene	PAO4 oil	0.01 wt.%	50%	greater than 50%	[79]
Graphene	Vegetable oil	-	From 0.0825 to 0.0714	From 414 to 374	[80]
GO	DI-H2O		26.1%	39.4%	[81]
GO	DI-H2O	0.1 wt.%	0.03 (2.3 times lower than 0.5 wt% ND)	-	[82]
GO	DI-H2O	0.1 wt.%	COF: 0.05	No obvious wear after 60 000 cycles	[83]
CNT	Pure palm oil	2 wt.%	COF: 0.121 (APE-10)	-	[84]
MXene $Ti_3C_2T_X$	TC-W outboard engine oil	0.01 wt.%	14.5%	6.3%	Our work

References

1. Rasheed, A.K.; Khalid, M.; Nor, A.F.; Wong, W.Y.; Duolikun, T.; Natu, V.; Barsoum, M.W.; Leo, B.F.; Zaharin, H.A.; Ghazali, M.J. MXene-Graphene Hybrid Nanoflakes as Friction Modifiers for Outboard Engine Oil. In Proceedings of the IOP Conference Series: Materials Science and Engineering, Chennai, India, 25–26 September 2020; Volume 834, p. 012039. [CrossRef]
2. Anh Tran, T. Some Methods to Prevent the Wear of Piston-Cylinder When Using Low Sulphur Fuel Oil (LSFO) for All Ships Sailing on Emission Control Areas (ECAs). In *Diesel and Gasoline Engines*; IntechOpen: London, UK, 2020.
3. UNEP (United Nations Environment Programme) Marine Pollution. Available online: https://www.unep.org/explore-topics/oceans-seas/ (accessed on 16 May 2023).
4. Chen, C.; Saikawa, E.; Comer, B.; Mao, X.; Rutherford, D. Ship Emission Impacts on Air Quality and Human Health in the Pearl River Delta (PRD) Region, China, in 2015, with Projections to 2030. *GeoHealth* **2019**, *3*, 284–306. [CrossRef]
5. UN (United Nation) Transforming Our World: The 2030 Agenda for Sustainable Development. Available online: https://sdgs.un.org/2030agenda (accessed on 16 May 2023).
6. Ali, Z.A.A.A.; Takhakh, A.M.; Al-Waily, M. A Review of Use of Nanoparticle Additives in Lubricants to Improve Its Tribological Properties. *Mater. Today Proc.* **2022**, *52*, 1442–1450. [CrossRef]
7. Ng, B.Y.S.; Ong, H.C.; Lau, H.L.N.; Ishak, N.S.; Elfasakhany, A.; Lee, H.V. Production of Sustainable Two-Stroke Engine Biolubricant Ester Base Oil from Palm Fatty Acid Distillate. *Ind. Crops Prod.* **2022**, *175*, 114224. [CrossRef]
8. Zaharin, H.A.; Ghazali, M.J.; Thachnatharen, N.; Ezzah, F.; Walvekar, R.; Khalid, M. Progress in 2D Materials Based Nanolubricants: A Review. *FlatChem* **2023**, *38*, 100485. [CrossRef]
9. Ji, Z.; Zhang, L.; Xie, G.; Xu, W.; Guo, D.; Luo, J.; Prakash, B. Mechanical and Tribological Properties of Nanocomposites Incorporated with Two-Dimensional Materials. *Friction* **2020**, *8*, 813–846. [CrossRef]
10. Rosenkranz, A.; Liu, Y.; Yang, L.; Chen, L. *2D Nano-Materials beyond Graphene: From Synthesis to Tribological Studies*; Springer International Publishing: Berlin/Heidelberg, Germany, 2020; Volume 10, ISBN 0123456789.
11. Al-Janabi, A.S.; Hussin, M.; Abdullah, M.Z. Stability, Thermal Conductivity and Rheological Properties of Graphene and MWCNT in Nanolubricant Using Additive Surfactants. *Case Stud. Therm. Eng.* **2021**, *28*, 101607. [CrossRef]
12. Yang, L.; Mao, M.; Huang, J.N.; Ji, W. Enhancing the Thermal Conductivity of SAE 50 Engine Oil by Adding Zinc Oxide Nano-Powder: An Experimental Study. *Powder Technol.* **2019**, *356*, 335–341. [CrossRef]

13. Anantha Kumar, K.; Sandeep, N.; Samrat, S.P.; Ashwinkumar, G.P. Effect of Electromagnetic Induction on the Heat Transmission in Engine Oil-Based Hybrid Nano and Ferrofluids: A Nanotechnology Application. *Proc. Inst. Mech. Eng. Part E J. Process Mech. Eng.* **2022**. [CrossRef]
14. Manu, B.R.; Gupta, A.; Jayatissa, A.H. Tribological Properties of 2D Materials and Composites—A Review of Recent Advances. *Materials* **2021**, *14*, 1630. [CrossRef]
15. Thachnatharen, N.; Khalid, M.; Arulraj, A.; Sridewi, N. Tribological Performance of Hexagonal Boron Nitride (HBN) as Nano-Additives in Military Grade Diesel Engine Oil. *Mater. Today Proc.* **2022**, *50*, 70–73. [CrossRef]
16. Liu, Y.; Dong, Y.; Zhang, Y.; Liu, S.; Bai, Y. Effect of Different Preparation Processes on Tribological Properties of Graphene. *Nanomater. Nanotechnol.* **2020**, *10*, 1847980420946655. [CrossRef]
17. Yi, S.; Li, J.; Liu, Y.; Ge, X.; Zhang, J.; Luo, J. In-Situ Formation of Tribofilm with Ti3C2Tx MXene Nanoflakes Triggers Macroscale Superlubricity. *Tribol. Int.* **2021**, *154*, 106695. [CrossRef]
18. Ramanavicius, S.; Ramanavicius, A. Progress and Insights in the Application of MXenes as New 2D Nanomaterials Suitable for Biosensors and Biofuel Cell Design. *Int. J. Mol. Sci.* **2020**, *21*, 9224. [CrossRef] [PubMed]
19. Naguib, M.; Mochalin, V.N.; Barsoum, M.W.; Gogotsi, Y. 25th Anniversary Article: MXenes: A New Family of Two-Dimensional Materials. *Adv. Mater.* **2013**, *26*, 992–1005. [CrossRef]
20. Naguib, M.; Come, J.; Dyatkin, B.; Presser, V.; Taberna, P.-L.; Simon, P.; Barsoum, M.W.; Gogotsi, Y. MXene: A Promising Transition Metal Carbide Anode for Lithium-Ion Batteries. *Electrochem. Commun.* **2012**, *16*, 61–64. [CrossRef]
21. Anasori, B.; Lukatskaya, M.R.; Gogotsi, Y. 2D Metal Carbides and Nitrides (MXenes) for Energy Storage. *Nat. Rev. Mater.* **2017**, *2*, 16098. [CrossRef]
22. Rosenkranz, A.; Righi, M.C.; Sumant, A.V.; Anasori, B.; Mochalin, V.N. Perspectives of 2D MXene Tribology. *Adv. Mater.* **2023**, *35*, 2207757. [CrossRef] [PubMed]
23. Cai, M.; Yan, H.; Song, S.; He, D.; Lin, Q.; Li, W.; Fan, X.; Zhu, M. State-of-the-Art Progresses for Ti3C2Tx MXene Reinforced Polymer Composites in Corrosion and Tribology Aspects. *Adv. Colloid Interface Sci.* **2022**, *309*, 102790. [CrossRef]
24. Miao, X.; Li, Z.; Liu, S.; Wang, J.; Yang, S. MXenes in Tribology: Current Status and Perspectives. *Adv. Powder Mater.* **2023**, *2*, 100092. [CrossRef]
25. Wyatt, B.C.; Rosenkranz, A.; Anasori, B. 2D MXenes: Tunable Mechanical and Tribological Properties. *Adv. Mater.* **2021**, *33*, 2007973. [CrossRef]
26. Zhou, Y.; Liu, M.; Wang, Y.; Yuan, J.; Men, X. Significance of Constructed MXene@Ag Hybrids for Enhancing the Mechanical and Tribological Performance of Epoxy Composites. *Tribol. Int.* **2022**, *165*, 107328. [CrossRef]
27. Parra-Muñoz, N.; Soler, M.; Rosenkranz, A. Covalent Functionalization of MXenes for Tribological Purposes—A Critical Review. *Adv. Colloid Interface Sci.* **2022**, *309*, 102792. [CrossRef] [PubMed]
28. Alhabeb, M.; Maleski, K.; Anasori, B.; Lelyukh, P.; Clark, L.; Sin, S.; Gogotsi, Y. Guidelines for Synthesis and Processing of Two-Dimensional Titanium Carbide ($Ti_3C_2T_x$ MXene). *Chem. Mater.* **2017**, *29*, 7633–7644. [CrossRef]
29. Naguib, M.; Presser, V.; Lane, N.; Tallman, D.; Gogotsi, Y.; Lu, J.; Hultman, L.; Barsoum, M.W. Synthesis of a New Nanocrystalline Titanium Aluminum Fluoride Phase by Reaction of Ti2AlC with Hydrofluoric Acid. *RSC Adv.* **2011**, *1*, 1493. [CrossRef]
30. Li, L.; Li, G.; Tan, L.; Zhang, Y.; Wu, B. Highly Efficiently Delaminated Single-Layered MXene Nanosheets with Large Lateral Size. *Langmuir* **2017**, *33*, 9000–9006. [CrossRef]
31. Wang, L.; Zhang, H.; Wang, B.; Shen, C.; Zhang, C.; Hu, Q.; Zhou, A.; Liu, B. Synthesis and Electrochemical Performance of Ti3C2Tx with Hydrothermal Process. *Electron. Mater. Lett.* **2016**, *12*, 702–710. [CrossRef]
32. Peng, C.; Wei, P.; Chen, X.; Zhang, Y.; Zhu, F.; Cao, Y.; Wang, H.; Yu, H.; Peng, F. A Hydrothermal Etching Route to Synthesis of 2D MXene (Ti3C2, Nb2C): Enhanced Exfoliation and Improved Adsorption Performance. *Ceram. Int.* **2018**, *44*, 18886–18893. [CrossRef]
33. Yang, S.; Zhang, P.; Wang, F.; Ricciardulli, A.G.; Lohe, M.R.; Blom, P.W.M.; Feng, X. Fluoride-Free Synthesis of Two-Dimensional Titanium Carbide (MXene) Using A Binary Aqueous System. *Angew. Chem.—Int. Ed.* **2018**, *57*, 15491–15495. [CrossRef]
34. Numan, A.; Rafique, S.; Khalid, M.; Zaharin, H.A.; Radwan, A.; Mokri, N.A.; Ching, O.P.; Walvekar, R. Microwave-Assisted Rapid MAX Phase Etching and Delamination: A Paradigm Shift in MXene Synthesis. *Mater. Chem. Phys.* **2022**, *288*, 126429. [CrossRef]
35. Gerard, O.; Numan, A.; Krishnan, S.; Khalid, M.; Subramaniam, R.; Kasi, R. A Review on the Recent Advances in Binder-Free Electrodes for Electrochemical Energy Storage Application. *J. Energy Storage* **2022**, *50*, 104283. [CrossRef]
36. Gao, F.; Lu, Q.; Komarneni, S. Fast Synthesis of Cerium Oxide Nanoparticles and Nanorods. *J. Nanosci. Nanotechnol.* **2006**, *6*, 3812–3819. [CrossRef] [PubMed]
37. Rosa, R.; Ponzoni, C.; Leonelli, C. Direct Energy Supply to the Reaction Mixture during Microwave-Assisted Hydrothermal and Combustion Synthesis of Inorganic Materials. *Inorganics* **2014**, *2*, 191–210. [CrossRef]
38. Nageswara Rao, B.; Ramesh Kumar, P.; Padmaraj, O.; Venkateswarlu, M.; Satyanarayana, N. Rapid Microwave Assisted Hydrothermal Synthesis of Porous α-Fe 2 O 3 Nanostructures as Stable and High Capacity Negative Electrode for Lithium and Sodium Ion Batteries. *RSC Adv.* **2015**, *5*, 34761–34768. [CrossRef]
39. Agusu, L.; Ahmad, L.O.; Anggara, D.; Alimin; Mitsudo, S.; Fujii, Y.; Kikuchi, H. Microwave Hydrothermal Synthesis of Reduced Graphene Oxide: Effects of Microwave Power and Irradiation Time. *J. Phys. Conf. Ser.* **2018**, *1011*, 012012. [CrossRef]
40. Kashinath, L. Microwave-Hydrothermal Synthesis of Copper Sulphide Nanorods Embedded on Graphene Sheets as an Efficient Electrocatalyst for Excellent Hydrogen Evolution Reaction. *Fuel* **2021**, *291*, 120143. [CrossRef]

41. Nagarajan, T.; Mohammad, K.; Nanthini, S.; Priyanka, J.; Rashmi, W. Microwave Synthesis of Molybdenum Disulfide Nanoparticles Using Response Surface Methodology for Tribological Application. *Nanomaterials* **2022**, *12*, 3369. [CrossRef]
42. Pan, X.; Shinde, N.M.; Lee, M.; Kim, D.; Kim, K.H.; Kang, M. Controlled Nanosheet Morphology of Titanium Carbide Ti3C2Tx MXene via Drying Methods and Its Electrochemical Analysis. *J. Solid State Electrochem.* **2020**, *24*, 675–686. [CrossRef]
43. Ghidiu, M.; Lukatskaya, M.R.; Zhao, M.-Q.; Gogotsi, Y.; Barsoum, M.W. Conductive Two-Dimensional Titanium Carbide 'Clay' with High Volumetric Capacitance. *Nature* **2014**, *516*, 78–81. [CrossRef]
44. Said, Z.; Sundar, L.S.; Tiwari, A.K.; Ali, H.M.; Sheikholeslami, M.; Bellos, E.; Babar, H. Recent Advances on the Fundamental Physical Phenomena behind Stability, Dynamic Motion, Thermophysical Properties, Heat Transport, Applications, and Challenges of Nanofluids. *Phys. Rep.* **2022**, *946*, 1–94. [CrossRef]
45. Kamarulzaman, M.K.; Hisham, S.; Kadirgama, K.; Ramasamy, D.; Samykano, M.; Said, Z.; Pandey, A.K. Improvement in Stability and Thermophysical Properties of CNC-MXene Nanolubricant for Tribology Application. *J. Mol. Liq.* **2023**, *381*, 121695. [CrossRef]
46. Mukherjee, S. Preparation and Stability of Nanofluids-A Review. *IOSR J. Mech. Civ. Eng.* **2013**, *9*, 63–69. [CrossRef]
47. Markandan, K.; Nagarajan, T.; Walvekar, R.; Chaudhary, V.; Khalid, M. Enhanced Tribological Behaviour of Hybrid MoS2@Ti3C2 MXene as an Effective Anti-Friction Additive in Gasoline Engine Oil. *Lubricants* **2023**, *11*, 47. [CrossRef]
48. Bao, Z.; Bing, N.; Zhu, X.; Xie, H.; Yu, W. Ti3C2Tx MXene Contained Nanofluids with High Thermal Conductivity, Super Colloidal Stability and Low Viscosity. *Chem. Eng. J.* **2021**, *406*, 126390. [CrossRef]
49. Nagarajan, T.; Khalid, M.; Sridewi, N.; Jagadish, P.; Shahabuddin, S.; Muthoosamy, K.; Walvekar, R. Tribological, Oxidation and Thermal Conductivity Studies of Microwave Synthesised Molybdenum Disulfide (MoS2) Nanoparticles as Nano-Additives in Diesel Based Engine Oil. *Sci. Rep.* **2022**, *12*, 14108. [CrossRef]
50. Guo, J.; Wu, P.; Zeng, C.; Wu, W.; Zhao, X.; Liu, G.; Zhou, F.; Liu, W. Fluoropolymer Grafted Ti3C2Tx MXene as an Efficient Lubricant Additive for Fluorine-Containing Lubricating Oil. *Tribol. Int.* **2022**, *170*, 107500. [CrossRef]
51. Liu, L.; Zhou, M.; Jin, L.; Li, L.; Mo, Y.; Su, G.; Li, X.; Zhu, H.; Tian, Y. Recent Advances in Friction and Lubrication of Graphene and Other 2D Materials: Mechanisms and Applications. *Friction* **2019**, *7*, 199–216. [CrossRef]
52. Suresha, B.; Hemanth, G.; Rakesh, A.; Adarsh, K.M. Tribological Behaviour of Pongamia Oil as Lubricant with and without Halloysite Nanotubes Using Four-Ball Tester. In Proceedings of the International Conference on Materials, Manufacturing and Machining 2019, Tamilnadu, India, 8–9 March 2019; p. 030011. [CrossRef]
53. Fan, W.; Zhu, X.; Ke, F.; Chen, Y.; Dong, K.; Ji, J.; Chen, B.; Tongay, S.; Ager, J.W.; Liu, K.; et al. Vibrational Spectrum Renormalisation by Enforced Coupling across the van Der Waals Gap between MoS2 and WS2 Monolayers. *Phys. Rev. B* **2015**, *92*, 24408. [CrossRef]
54. Rosenkranz, A.; Grützmacher, P.G.; Espinoza, R.; Fuenzalida, V.M.; Blanco, E.; Escalona, N.; Gracia, F.J.; Villarroel, R.; Guo, L.; Kang, R.; et al. Multi-Layer Ti3C2Tx-Nanoparticles (MXenes) as Solid Lubricants—Role of Surface Terminations and Intercalated Water. *Appl. Surf. Sci.* **2019**, *494*, 13–21. [CrossRef]
55. Zhang, X.; Guo, Y.; Li, Y.; Liu, Y.; Dong, S. Preparation and Tribological Properties of Potassium Titanate-Ti3C2Tx Nanocomposites as Additives in Base Oil. *Chin. Chem. Lett.* **2019**, *30*, 502–504. [CrossRef]
56. Jin, W.; Jiang, L.; Han, L.; Huang, H.; Zhang, J.; Guo, M.; Gu, Y.; Zhi, F.; Chen, Z.; Yang, G. Investigation of Thermal Conductivity Enhancement of Water-Based Graphene and Graphene/MXene Nanofluids. *J. Mol. Liq.* **2022**, *367*, 120455. [CrossRef]
57. Walvekar, R.; Faris, I.A.; Khalid, M. Thermal Conductivity of Carbon Nanotube Nanofluid-Experimental and Theoretical Study. *Heat Transf.—Asian Res.* **2012**, *41*, 145–163. [CrossRef]
58. Baby, T.T.; Ramaprabhu, S. Investigation of Thermal and Electrical Conductivity of Graphene Based Nanofluids. *J. Appl. Phys.* **2010**, *108*, 124308. [CrossRef]
59. Aslfattahi, N.; Samylingam, L.; Abdelrazik, A.S.; Arifutzzaman, A.; Saidur, R. MXene Based New Class of Silicone Oil Nanofluids for the Performance Improvement of Concentrated Photovoltaic Thermal Collector. *Sol. Energy Mater. Sol. Cells* **2020**, *211*, 110526. [CrossRef]
60. Qing, S.H.; Rashmi, W.; Khalid, M.; Gupta, T.C.S.M.; Nabipoor, M.; Hajibeigy, M.T. Thermal Conductivity and Electrical Properties of Hybrid SiO 2 -Graphene Naphthenic Mineral Oil Nanofluid as Potential Transformer Oil. *Mater. Res. Express* **2017**, *4*, 015504. [CrossRef]
61. Yiamsawas, T.; Dalkilic, A.S.; Mahian, O.; Wongwises, S. Measurement and Correlation of the Viscosity of Water-Based Al2O3 and TiO2 Nanofluids in High Temperatures and Comparisons with Literature Reports. *J. Dispers. Sci. Technol.* **2013**, *34*, 1697–1703. [CrossRef]
62. Malekzadeh, A.; Pouranfard, A.R.; Hatami, N.; Kazemnejad Banari, A.; Rahimi, M.R. Experimental Investigations on the Viscosity of Magnetic Nanofluids under the Influence of Temperature, Volume Fractions of Nanoparticles and External Magnetic Field. *J. Appl. Fluid Mech.* **2016**, *9*, 693–697. [CrossRef]
63. Apmann, K.; Fulmer, R.; Soto, A.; Vafaei, S. Thermal Conductivity and Viscosity: Review and Optimisation of Effects of Nanoparticles. *Materials* **2021**, *14*, 1291. [CrossRef] [PubMed]
64. Chen, J.; Zhao, W. Simple Method for Preparing Nanometer Thick Ti3C2TX Sheets towards Highly Efficient Lubrication and Wear Resistance. *Tribol. Int.* **2021**, *153*, 106598. [CrossRef]
65. Nguyen, H.T.; Chung, K.-H. Assessment of Tribological Properties of Ti3C2 as a Water-Based Lubricant Additive. *Materials* **2020**, *13*, 5545. [CrossRef]
66. Liu, Y. Synthesis and Tribological Property of Ti3C2TX Nanosheets. *J. Mater. Sci.* **2016**, *52*, 2200–2209. [CrossRef]

67. Xue, M.; Wang, Z.; Yuan, F.; Zhang, X.; Wei, W.; Tang, H.; Li, C. Preparation of TiO2/Ti3C2Tx Hybrid Nanocomposites and Their Tribological Properties as Base Oil Lubricant Additives. *RSC Adv.* **2017**, *7*, 4312–4319. [CrossRef]
68. Zhang, X.; Xue, M.; Yang, X.; Wang, Z.; Luo, G.; Huang, Z.; Sui, X.; Li, C. Preparation and Tribological Properties of Ti3C2(OH)2 Nanosheets as Additives in Base Oil. *RSC Adv.* **2015**, *5*, 2762–2767. [CrossRef]
69. Yang, J.; Chen, B.; Song, H.; Tang, H.; Li, C. Synthesis, Characterization, and Tribological Properties of Two-Dimensional Ti3C2. *Cryst. Res. Technol.* **2014**, *49*, 926–932. [CrossRef]
70. Feng, Q.; Deng, F.; Li, K.; Dou, M.; Zou, S.; Huang, F. Enhancing the Tribological Performance of Ti3C2 MXene Modified with Tetradecylphosphonic Acid. *Colloids Surf. A Physicochem. Eng. Asp.* **2021**, *625*, 126903. [CrossRef]
71. Gao, J.; Du, C.-F.; Zhang, T.; Zhang, X.; Ye, Q.; Liu, S.; Liu, W. Dialkyl Dithiophosphate-Functionalized Ti3C2Tx MXene Nanosheets as Effective Lubricant Additives for Antiwear and Friction Reduction. *ACS Appl. Nano Mater.* **2021**, *4*, 11080–11087. [CrossRef]
72. Zhang, F.X.; Su, X.; Tang, G.G.; Xu, J. Construction and Tribological Behaviors of MXenes/MoS2 Heterojunction with 2D/2D Structure in Liquid Paraffin. *Chalcogenide Lett.* **2021**, *18*, 225–235. [CrossRef]
73. Feng, P.; Ren, Y.; Li, Y.; He, J.; Zhao, Z.; Ma, X.; Fan, X.; Zhu, M. Synergistic Lubrication of Few-Layer Ti3C2Tx/MoS2 Heterojunction as a Lubricant Additive. *Friction* **2022**, *10*, 2018–2032. [CrossRef]
74. Zhou, C.; Li, Z.; Liu, S.; Ma, L.; Zhan, T.; Wang, J. Synthesis of MXene-Based Self-Dispersing Additives for Enhanced Tribological Properties. *Tribol. Lett.* **2022**, *70*, 63. [CrossRef]
75. Jazaa, Y.; Lan, T.; Padalkar, S.; Sundararajan, S. The Effect of Agglomeration Reduction on the Tribological Behavior of WS2 and MoS2 Nanoparticle Additives in the Boundary Lubrication Regime. *Lubricants* **2018**, *6*, 106. [CrossRef]
76. Guo, Y.B.; Zhang, S.W. The Tribological Properties of Multi-Layered Graphene as Additives of PAO2 Oil in Steel-Steel Contacts. *Lubricants* **2016**, *4*, 30. [CrossRef]
77. Azman, S.S.N.; Zulkifli, N.W.M.; Masjuki, H.; Gulzar, M.; Zahid, R. Study of Tribological Properties of Lubricating Oil Blend Added with Graphene Nanoplatelets. *J. Mater. Res.* **2016**, *31*, 1932–1938. [CrossRef]
78. Cai, Z.B.; Zhao, L.; Zhang, X.; Yue, W.; Zhu, M.H. Combined Effect of Textured Patterns and Graphene Flake Additives on Tribological Behavior under Boundary Lubrication. *PLoS ONE* **2016**, *11*, e0152143. [CrossRef] [PubMed]
79. Zheng, D.; Cai, Z.B.; Shen, M.X.; Li, Z.Y.; Zhu, M.H. Investigation of the Tribology Behaviour of the Graphene Nanosheets as Oil Additives on Textured Alloy Cast Iron Surface. *Appl. Surf. Sci.* **2016**, *387*, 66–75. [CrossRef]
80. Kiu, S.S.K.; Yusup, S.; Soon, C.V.; Arpin, T.; Samion, S.; Kamil, R.N.M. Tribological Investigation of Graphene as Lubricant Additive in Vegetable Oil. *J. Phys. Sci.* **2017**, *28*, 257–267. [CrossRef]
81. Li, X.; Liu, Q.; Dong, G. Self-Assembly Membrane on Textured Surface for Enhancing Lubricity of Graphene Oxide Nano-Additive. *Appl. Surf. Sci.* **2020**, *505*, 144572. [CrossRef]
82. Wu, P.; Chen, X.; Zhang, C.; Luo, J. Synergistic Tribological Behaviors of Graphene Oxide and Nanodiamond as Lubricating Additives in Water. *Tribol. Int.* **2019**, *132*, 177–184. [CrossRef]
83. Yang, J.; Xia, Y.; Song, H.; Chen, B.; Zhang, Z. Synthesis of the Liquid-like Graphene with Excellent Tribological Properties. *Tribol. Int.* **2017**, *105*, 118–124. [CrossRef]
84. Gao, T.; Li, C.; Zhang, Y.; Yang, M.; Jia, D.; Jin, T.; Hou, Y.; Li, R. Dispersing Mechanism and Tribological Performance of Vegetable Oil-Based CNT Nanofluids with Different Surfactants. *Tribol. Int.* **2019**, *131*, 51–63. [CrossRef]

lubricants

Article

Effect of Cu-Doped Carbon Quantum Dot Dispersion Liquid on the Lubrication Performance of Polyethylene Glycol

Shusheng Liu [1], Xiuqian Yu [1], Enzhu Hu [1,*], Enhao Su [1], Yanjie Chen [1], Jianping Wang [1], Kunhong Hu [1], Yong Xu [1], Xianguo Hu [2,*] and Hua Zhong [1]

[1] School of Energy Materials and Chemical Engineering, Hefei University, Hefei 230601, China
[2] Institute of Tribology, Hefei University of Technology, Hefei 230009, China
* Correspondence: huez@hfuu.edu.cn (E.H.); xghu@hfut.edu.cn (X.H.)

Abstract: Energy saving and reduced consumption of key materials such as bearings in high-end equipment can be realized by synthesizing a new lubricating functional additive, copper-doped carbon quantum dot dispersion liquid (Cu-CQDs) via hydrothermal reaction with glycerol, cupric chloride dihydrate, and choline chloride as raw materials. The influence of the dispersion liquid containing Cu-CQDs nanoparticles on the lubricating properties of polyethylene glycol (PEG200) was investigated on a four-ball friction tester. The wear scars of steel balls after friction were analyzed using a scanning electron microscope accompanied by energy dispersive spectroscopy (SEM/EDS), photoelectron microscopy, and Raman spectroscopy. The results revealed the friction and wear mechanism of Cu-CQDs. Cu-CQDs dispersion liquid can significantly enhance the lubrication performance of PEG. The average friction coefficient of PEG containing 2.0 wt% Cu-CQDs dispersion liquid was 40.99% lower than that of pure PEG. The friction and wear mechanism can be ascribed to friction, inducing Cu-CQDs to participate in the formation of boundary lubricating film, resulting in a low friction coefficient and wear scar diameter.

Keywords: copper doping; carbon quantum dots; boundary lubricating film; friction and wear mechanism

1. Introduction

Nanoparticles have broad application prospects because of their unique nano-size effect and superior chemical properties [1]. Many kinds of nano-materials exist, including copper, silver, nickel, molybdenum disulfide, tungsten disulfide, graphite nanosheets, nanospheres, graphene, carbon nanotubes, and carbon quantum dots (CQDs), which can be divided into organic, inorganic, and metal-based nanoparticles. Nanoparticles have been widely used in real life as lubricant additives. Its excellent antifriction and anti-wear properties and unique self-healing ability are potential substitutes for traditional additives that meet the developing requirements of high-performance lubricants [2,3]. It is well known that a large amount of energy loss in industrial production is caused by friction and wear. Improved lubrication can save 30% of energy. Therefore, how to save energy and reduce consumption is the focus of tribology scientists. The use of superior lubricants with the proper functional additives can prolong the life of mechanical components and realize energy savings and reduced consumption [4,5]. He et al. [6] used it as a functional additive of N32 lubricating oil. The nano copper bearing function improved the lubricating film's wear resistance and played a good role in reducing friction and anti-wear. Qin et al. [7] investigated halloysite nanotubes as an oil-based additive to improve the anti-wear properties. The outstanding wear resistance of the surface lubricated by the calcined product is attributed to the faster and easier formation of a multilayer nanocrystalline tribo-film. Jiang et al. [8] investigated and compared the tribological behaviour of steel/steel surface lubricated by poly alpha olefin (PAO 6) oil containing MoS_2/h-BN hybrid nanoparticles. The formation of the tribo-film on the surface of the tribo-pair resulted in a lower friction coefficient and wear rate. Lahouij et al. [9] have studied the influence of the crystal structure of IF-MoS_2 nanoparticles on the lubricating properties of the fullerenes added to a synthetic base oil. The good lubricating properties of poorly crystallized particles by their higher ability to exfoliate and to form rapidly a tribo-film made of h-MoS_2 sheets on the surfaces compared to the perfectly crystallized particles. Li et al. [10] reported that nano-Cu powder can significantly improve the lubrication performance of PEG. The tribological mechanism can be attributed to the formation of copper deposition film by nano-Cu, thereby reducing the friction coefficient and the wear of friction pairs.

As a new material in the carbon family, CQDs are usually mono-disperse spherical carbon nanomaterials with a size of less than 10 nm, which are new nanomaterials with various industrial application prospects [11,12]. The research on the tribology of CQDs is still in its infancy. CQDs have remarkable tribological properties and great potential to become green and efficient lubricating materials. Kumar et al. [13] studied and introduced several applications, development progress, antifriction performance, anti-wear performance, and lubrication mechanism of lubricant additives based on carbon quantum dots (CQDs), and provided some suggestions on complete functionalized or mixed doped CQDs materials. Tang et al. [14] made a comprehensive and systematic review of the research progress of carbon dots in the lubrication field. The research work of carbon dots in the field of lubricating functional additives and lubricating coatings has been described in detail. Three strategies for improving the lubricating effect (size and shape control, surface modification, and heteroatom doping) have been pointed out, and the lubricating mechanism of CQDs and the main challenges have been comprehensively analyzed. Ma et al. [15] studied that CQDs and ionic liquid (IL) nanoparticles prepared by simple modification showed excellent tribological properties as lubricating materials and obtained a super-low friction coefficient of about 0.006 and wear rate of about 0.7×10^{-14} m^3/Nm with the CQDs content of 3.6 wt%. The results show that IL-modified CQDs can provide a new idea for studying various lubricating materials to achieve ultra-low friction and can further study other practical applications. Shang et al. [16] studied chemically grafted CQDs and IL hybrid nanomaterials (CQDs/IL), and the hybrid nanomaterials with spherical shell structure showed a synergistic effect in improving the friction reduction and anti-wear properties of boundary lubrication. Tang et al. [17] found that the introduction of 0.1 wt% CQDs in water reduced the friction coefficient and wear rate by 33% and 80%, respectively. This finding can be attributed to the excellent tribological energy of CQDs from minimizing CQD sediment particles and reducing friction and wear through nano-bearing and nano-filling mechanisms. Hu et al. [18] prepared water-soluble CQDs by adjusting the carbonation degree of ammonium citrate through a simple, low-cost bottom-up method. The lubricity and inhibition of CQDs as nano-additive of water-based lubricant were studied. CQDs as environment-friendly nano-additives have low friction coefficients, excellent wear resistance, and excellent inhibition effect.

Limited studies have focused on the tribological properties of element-doped or hybrid CQDs. He et al. [19] prepared a new type of nitrogen-doped CQD (N-CQD) nanoparticles by solvothermal method. The particle diameter is about 10 nm. Nanoparticles have a graphene oxide (GO) micro graphitization structure, and they were added to MoS_2 nanofluids. The tribological properties of the composite fluid were investigated, and the results show that the friction process of MoS_2 and N-CQDs nanofluids was in a mixed lubrication state. The particles do not interfere with each other and play a synergistic role in lubrication. Ye et al. [20] studied nitrogen-doped CQDs (N-CQDs) as functional additives and explored their influence on the lubricating properties of bio-based castor oil. The chemical grafting of CQDs in the olefins (PAO) with good dispersion. The friction mechanism can be attributed to the continuous deposition of the wear surface by nanoparticles at the interface of the friction pair, forming a thin protective film containing iron oxide, inorganic carbon, and nitrogen. Tu et al. [21] synthesized a new green lubricating oil additive CQDs particle-doped nickel (Ni-CQDs) with citric acid and nickel acetate as raw materials. The effects of CQDs and Ni-CQDs nanoparticles on the friction behaviour of polyethylene glycol (PEG200) were investigated. Results show that both CQDs and Ni-CQDs particles can enhance the lubrication performance of PEG200. Moreover, Ni-CQDs nanoparticles can enhance the lubricating performance compared with ordinary CQD particles. This finding can be attributed to the fact that friction induces Ni-CQDs to participate in the formation of friction film, resulting in a low friction coefficient and wear rate. Huang et al. [22] found that when CQDs/CuSx nanocomposite particles are used as additives, they show excellent lubrication and superior repair ability of worn surfaces. The mechanism can be attributed to the combination of the multilayer graphite structure of CQDs and the high chemical activity of CuSx. Shang et al. [23] prepared a hybrid of CQDs and graphene oxide (GO) through a one-pot method. The tribological behaviour of PEG as a lubricating additive was studied. A significant antifriction effect was observed, and the mechanism can be attributed to the synergistic effect of adsorption of spherical CQDs and GOs on the sliding surface, which effectively protected the interface from bad friction and wear.

Limited research has focused on the preparation of metal-based biomass CQDs dispersion and its tribological properties. Therefore, the copper-doped CQDs dispersion was prepared using a traditional hydrothermal synthesis method, and the influence of different concentrations of the dispersion on the lubrication characteristics of PEG lubricant was investigated. The successful implementation of this work has laid a certain foundation for the development of high-performance lubricants and provided basic data for enriching the tribological research of metal-based CQDs.

2. Experimental

2.1. Preparation and Dispersion of Materials

2.1.1. Preparation of Copper-Doped CQDs Dispersion

Precursor solution (DES): Choline chloride (analytically pure, Shanghai Zhanyun Chemical Co., Ltd., Shanghai, China), cupric chloride dihydrate (analytically pure, Shanghai Zhanyun Chemical Co., Ltd.), and glycerol (analytically pure, Sinopharm Chemical Reagent Co., Ltd., Shanghai, China) were weighed at a proportion of 5:1:1, and the mixed solution was poured into a 50 mL three-necked flask and placed in an oil bath to prevent liquid splashing and water loss during heating. The bottle mouth was covered with a rubber stopper. At room temperature, the oil bath was stirred and heated to 110 °C, and the reaction lasted for 5 h until the precursor solution became stable. After the solution was cooled to room temperature, it was sealed for use.

Hydrothermal synthesis: Approximately 1.0 g of DES solution was obtained, mixed with 30 mL of pure water, placed in a beaker, and sealed. The mixture was stirred with small magnets for 2 h until the DES solution was uniformly dispersed in pure water. The sample was placed in mixed DES solution into a 50 mL reaction kettle lined with polytetrafluoroethylene, the shell was tightened, the oven temperature was set to 220 °C, and it was held for 24 h for the successful preparation of Cu-CQDs. After the reaction, 0.22 μ the supernatant was filtered out by PVDF membrane, and the macromolecular impurities were removed by dialysis with 3500D dialysis bag for 2 days. After rotary evaporation, the solution was freeze-dried to obtain copper-doped carbon dots dispersion.

2.1.2. The Mutual Solubility of Copper-Doped CQDs Dispersion with PEG200

The dispersion containing copper-doped carbon nanoparticles was added to the base oil PEG200 (99%, Shandong Yousuo Chemical Co., Ltd., Heze, China) as a lubricating additive. The dispersion characteristics of its nanoparticle dispersion with different mass fractions in PEG are shown in Figure 1. The figure that copper-doped carbon dots have good dispersion in PEG. The yield of copper-doped carbon nanoparticle dispersion is approximately 95.35%. The formation mechanism can be summarized as follows: during the formation of doped carbon dots, glycerol molecules, and cupric chloride molecules experienced a process of "Dispersion pyrolysis, Aggregation carbonization, Hydrothermal recombination" three kinds of reaction processes [24]. The specific colour development mechanism can be expressed in publicity (1).

$$[Cu(H_2O)_4]^{2+} + 4Cl^- \leftrightharpoons (CuCl_4)^{2-} + 4H_2O, H > 0 \quad (1)$$

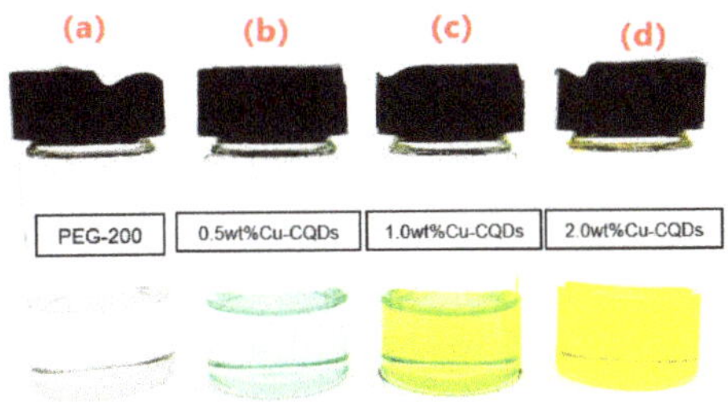

Figure 1. Mutual solubility of Cu-CQDs dispersions with different mass concentrations in PEG200.

Cu-CQD dispersions with different amounts of addition show different colours in PEG. The colour change mechanism can be explained as follows: After dissolution in PEG200, hydrated copper ions $[Cu(H_2O)_4]^{2+}$, and tetrachloride copper complex ions $[CuCl_4]^{2-}$ are formed by the combination of Cl^- and copper ions. The colour of the ions is yellow, because the concentration of Cu-CQDs changes the equilibrium constant of copper tetrachloride ion, making it move forward. With the increase in concentration, the colour of the solution changed from blue to yellow-green and then yellow (Figure 1 shows the colour change of copper-based CQDs in PEG).

2.2. *Characterization of Physical and Chemical Properties of Cu-CQDs Particles in Dispersion Solution*

Transmission electron microscopy (TEM, FEI Tecnai 200 kV) was used to characterize the internal structure of the Cu-CQDs particles in the dispersion solution and the morphological characteristics of the primary particles. As shown in Figure 2a, the Cu-CQDs particles were uniformly dispersed, and the average particle size was approximately 8.33 nm. Then, a fluorescence spectrophotometer (Edinburgh FLS1000, EI) was used to obtain the fluorescence emission spectrum of copper-based carbon dots at the excitation wavelength of 360 nm and at room temperature. Figure S1 shows that the maximum emission wavelength of fluorescence is 525 nm, and the maximum emission wavelength does not change with time without a blue shift or red shift. The absorption spectrum of CQDs was measured using an ultraviolet–visible absorption spectrometer (UV-1800, Mapada). The dispersion of Cu-CQDs in water was conducted with Zeta spectroscopy (model Zs90). It indicated that Cu-CQDs particles could be dispersed uniformly as shown in Figure 2b,c shows that their absorption ranges are all in the ultraviolet light region, and relatively wide absorption peaks belonging to the near ultraviolet region were observed between 200–250 nm, which can be attributed to π-π * transition of aromatic sp2 hybridization. The molecular structure of the material was characterized using a Fourier transform infrared spectrometer (NicoletiS50). The functional group absorption peak was inferred based on the vibration form of the molecular group corresponding to the absorption peak. As shown in Figure S2, the peak at 3357 cm^{-1} of Cu-CQDs shown in Figure 2b is the absorption peak of -OH, the peak at 1636 cm^{-1} is the absorption peak of C=O, and the peak at 1080 cm^{-1} is the characteristic absorption peak of C-O stretching vibration in -C-OH [25,26]. PEG200 is a hydrophilic polymer with a large number of hydroxyl groups on its molecular surface, which makes it show polar properties and can be well compatible with Cu-CQDs, which is also the reason why PEG is selected as the base oil. Figure S2c shows the infrared spectrum of PEG200. The peak at 3405 cm^{-1} is the absorption peak of -OH. In the figure, there is a strong absorption peak at 1062 cm^{-1}, which is the characteristic absorption peak of C-O-C stretching vibration in the PEG200 molecule. Figure 2a shows the infrared spectrum of the added amount of 1.0 wt% Cu-CQDs in PEG200, the functional groups are almost the same as PEG due to the small proportion of the added amount.

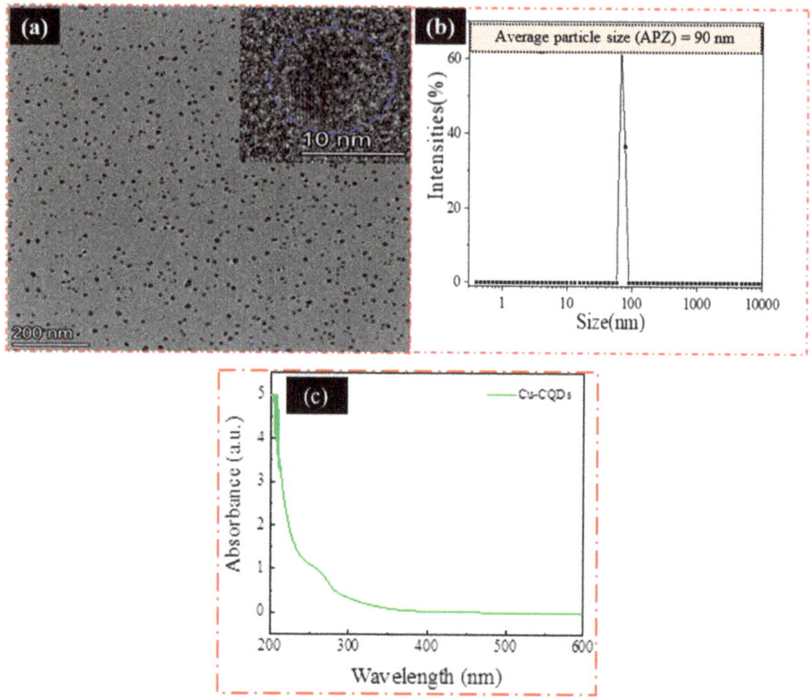

Figure 2. HRTEM, particle size distribution of Cu-CQDs in water and UV absorption spectrum of copper-doped carbon quantum dots. (**a**) TEM images of Cu-CQDs; (**b**) particle size distribution of Cu-CQDs in water; and (**c**) ultraviolet visible absorption spectrum.

2.3. Tribological Tests

The four-ball friction testing machine (MRS-10D type) is shown in Figure 3. Based on the petrochemical industry standard parameters of 392 N and 1200 rpm/min, the tribological test was carried out for 30 min at 25, 75, and 100 °C [27]. The GB/3142-2019 test standard was used to test the maximum seizure load (P_B value, accuracy less than 10 N) of solution oils with different concentrations of Cu-CQDs. Table 1 provides the P_B values of solution oils with different concentrations of Cu-CQDs. The four steel balls were arranged according to the regular tetrahedron, the upper ball rotated at 1450 $\pm$ 5 rpm/min, the lower three balls were fixed together with oil boxes, and the load was applied to the steel balls from bottom to top through the hydraulic system. During the test, the contact points of the four steel balls were immersed in the lubricant. Each test lasted for 10 s. After the test, the wear spot diameter of any steel ball in the oil tank was measured. The test was repeated according to the specified procedure until the evaluation index representing the bearing capacity of the lubricant was obtained [28].

The steel ball used in the test was ASTM E2100 bearing steel with a diameter of 12.7 mm, which was manufactured according to the Chinese national standard GB/T308-2002. The hardness of steel was set to HRC = 65, and the average surface roughness was Ra = 0.012 μm [29]. The lubrication state was reduced to the boundary lubrication condition by analysis of the friction coefficients of samples and surface contact stress of steel balls [30]. The antifriction and wear behaviours of Cu-CQDs as a lubricating additive under different working conditions were explored by subjecting all oil samples to ultrasonic dispersion treatment for

30 min. The test steel balls and oil box abrasives were cleaned with ethanol or petroleum ether. Each group of experiments was carried out thrice independently to reduce the experimental deviation [31,32]. SEM/EDS and a 3D laser scanning microscope (VK-X100k) were used to analyze the morphology of the wear scar area and the element composition on the surface. The valence states of chemical elements were studied by X-ray photoelectron spectroscopy (ESCALAB 250), and the mechanism of friction and wear was expounded.

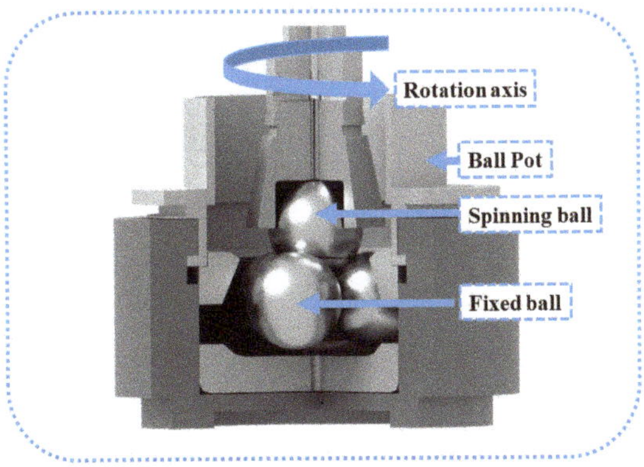

Figure 3. Schematic diagram of a four balls tribo-meter.

Table 1. P_B value of solution oil with different concentrations of Cu-CQDs.

Sample	Maximum Non-Seizure Load (N)	Rate of Change (%)
PEG200	530	-
0.5 wt% Cu-CQDs	588	10.94%
1.0 wt% Cu-CQDs	588	10.94%
2.0 wt% Cu-CQDs	637	20.19%

3. Results and Discussion

3.1. Antifriction Property

Figure 4 shows the influence of different amounts of dispersion on PEG lubrication characteristics at 25, 75, and 100 °C. As shown in Figure 4a, at 25 °C, with the increase in the content of Cu-CQD dispersion, the friction coefficient also decreased, with a maximum reduction of 19.83%, reflecting the Cu-CQDs can promote the antifriction property of PEG200. Figure 4b,c show the change in friction coefficient under the lubrication of different amounts of dispersion at 75 and 100 °C. With the increase in the content of Cu-CQD dispersion, the friction coefficient decreased. With the increase in temperature, the friction coefficient became considerably lower than that of pure PEG200. At 75 °C the highest reduction was 31.86%. At 100 °C, the maximum reduction was 40.99%. At high temperature, the lubricating performance of the oil sample was significantly enhanced, and the results are shown in Figure 4c,d. By using the formulas $\sigma = 1.5 \times W_1/(\pi \times a^2)$ [33], the contact surface stress between steel balls during lubrication (W_1 refers to the pressure between steel balls, and a refers to the circular contact area between steel balls) can be calculated, and the results are shown in Table 2. Combined with the friction coefficients analysis of samples, which indicates that Cu-CQDs are in a boundary lubrication state when lubricated with PEG at different temperatures, friction can induce Cu-CQDs to participate in the formation of boundary lubrication film.

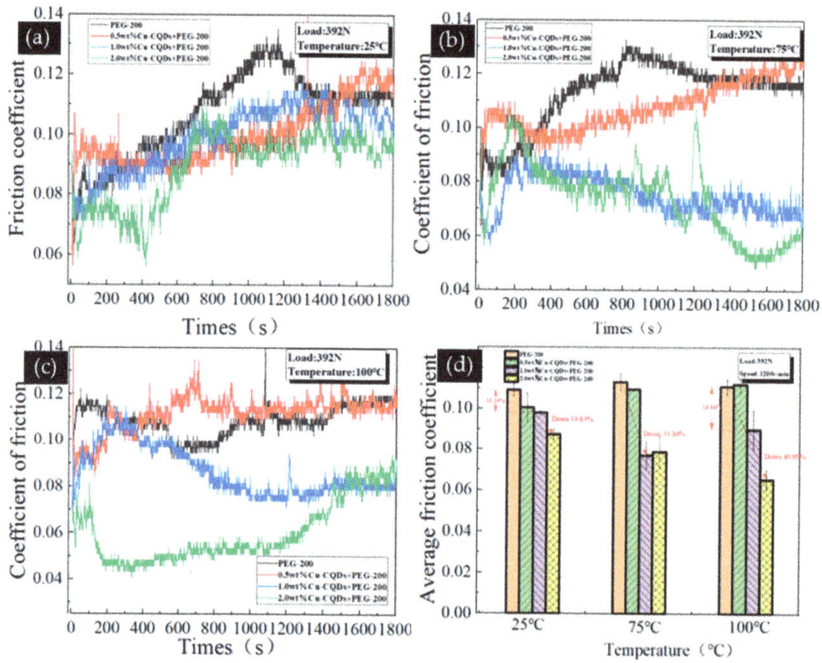

Figure 4. The changes of friction coefficients with different mass fractions of Cu-CQDs dispersion (1200 rpm/min, each group lasts for 30 min), (**a**) 25 °C; (**b**) 75 °C; (**c**) 100 °C; and (**d**) histogram of average friction coefficient under different working conditions.

Table 2. Contact surface stress of oil samples under different temperatures.

Samples	25 °C (Gpa)	75 °C (Gpa)	100 °C (Gpa)
PEG200	0.33	0.54	0.24
0.5 wt% Cu-CQDs	0.51	0.47	0.37
1.0 wt% Cu-CQDs	0.43	0.68	0.51
2.0 wt% Cu-CQDs	0.41	0.36	0.29

3.2. *Abrasion Resistance Property*

Figure 5 shows the changes in average wear scar diameter (AWSD) under oil sample lubrication with different Cu-CQD dispersions at different temperatures. When the oil temperature was 25 °C the AWSD diameter of pure PEG200 was 0.71 mm. When Cu-CQD dispersions with different contents were added, AWSD decreased slightly. The oil film thickness of PEG possibly took an important role in reducing the AWSDs. However, when adding high contents of Cu-CQDs, the aggregation role resulted in the high AWSDs.

At 75 °C, the AWSD of pure PEG200 was 0.56 mm. When 1.0 wt% Cu-CQDs dispersion was added, the AWSD of the oil sample decreased significantly. Friction induced the formation of tribo-film including components of Cu-CQDs particles. Chemical reactions took place on the surfaces of steel balls during the friction process which is better than that of oil film thickness. The AWSD of pure PEG was 0.83 mm at 100 °C. When 0.1 wt% Cu-CQDs dispersion was added, the AWSD decreased to 0.57 mm, with a reduction rate of 31.2%. Cu-CQDs nanoparticles can significantly improve the anti-wear properties of base oil PEG200 at higher temperatures. However, with the increase in oil temperature from room temperature to 75 and 100 °C, when 2.0 wt% of Cu-CQDs was added, AWSD is higher than that of other additions. After the tests, scratches of different degrees were found at the bottom of the steel balls. When a high concentration of Cu-CQDs was added to PEG200, a certain amount of Cu-CQDs nanoparticles or wear debris at the interface of the friction pair will agglomerate, and abrasive wear will occur at the interface of the friction pair, thus reducing the wear resistance property of PEG [34]. Friction film plays an important role in wear resistance. The optimal concentration of Cu-CQDs nanoparticles additive was 1.0 wt%. P_B value in Table 1 also confirms the above experimental results, and Cu-CQDs nanoparticles had the most significant wear resistance at high temperatures.

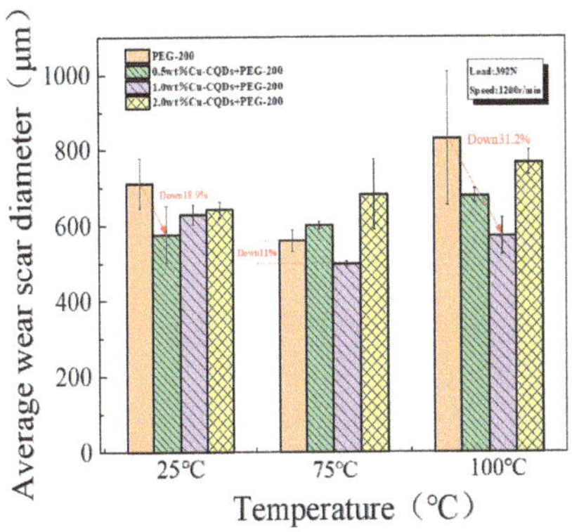

Figure 5. Change of average wear scar diameter under lubrication of Cu-CQDs dispersion with different mass fractions (1200 rpm/min, each group lasts for 30 min).

3.3. Surface Analysis

The micro-oil stain analysis system (YP-2 type) was used to analyze the oil stain after the experiment. Figure S3 shows the oil stain of 2.0 wt% Cu-CQDs at different temperatures. The oil stain was possibly composed of copper ion and chloride ion complexes, iron filings and iron oxides, and copper, and its derivatives. After friction, the wear debris and abrasive particles in the dispersion show agglomeration in Figure S3a, and the colour of the oil sample was significantly deepened in Figure S3c. Friction induced the oxidation of the oil.

Figures 6–8 show the cross-sectional area, three-dimensional morphology, average wear scar diameter, and its change rate of wear scar area lubricated by different oil samples at different temperatures. At 25 °C, many wide pits and deep grooves were observed on the surface of the steel ball, as indicated by the red arrow in Figure 6. This finding shows that the protection ability of pure PEG on the surface of steel balls is poor. When different amounts of Cu-CQDs were added, the width and depth of the furrow on the surface of the steel ball were significantly reduced, and the lubricating film formed. The average wear scar diameter on the steel ball surface decreased significantly. When 0.5 wt% Cu-CQDs was added, the AWSD decreased by 18.93% compared with pure PEG as shown in Figure 6. At 75 °C, under pure PEG lubrication, the surface furrow was very deep, but the average wear scar diameter slightly changed to approximately 560.72 μm. Different amounts of Cu-CQDs were added, and only 1 wt% Cu-CQDs reduced the average wear scar diameter by 11%, as shown in Figure 7. This conclusion can be attributed to the decomposition of PEG at 75 °C, resulting in the improvement of lubrication performance, and the lubrication efficiency is better than that of Cu-CQDs particles. At 100 °C, when 1.0 wt% Cu-CQDs was added to PEG, the wear spot diameter of the steel ball decreased by the largest proportion. In comparison with pure PEG lubrication (0.83 mm), the wear spot diameter decreased by 31.2%, as shown in Figure 8.

Figure 9 shows SEM/EDS analysis of the steel ball surface under different oil sample lubrication. Figure 9a shows a deep furrow on the surface of the steel ball lubricated by pure PEG, as indicated by the red arrow. This finding shows that PEG has a poor ability to protect the surface of the steel ball. The energy spectrum analysis shows that the main elements on the surface are carbon, oxygen, and iron. When 0.5 wt% Cu-CQDs were added to PEG, the width of the furrow of the ink mark on the steel ball surface was significantly reduced, and the friction-induced boundary film on the surface was formed as shown by the red arrow (Figure 9b). The energy spectrum analysis of the wear mark area shows that, except for the steel ball component elements, the copper content was significantly increased, reaching 0.8 wt%. Therefore, the friction-induced copper-doped carbon dots participated in the composition of the boundary film. The furrow depth on the surface of the steel ball was significantly reduced, and the surface smoothness was significantly better than that of the sample without carbon dots when the steel ball was continuously added at contents of 1.0 and 2.0 wt%. The results also confirmed that the copper-doped carbon dots participated in the formation of boundary lubrication film and remarkably improved the lubrication characteristics of PEG200.

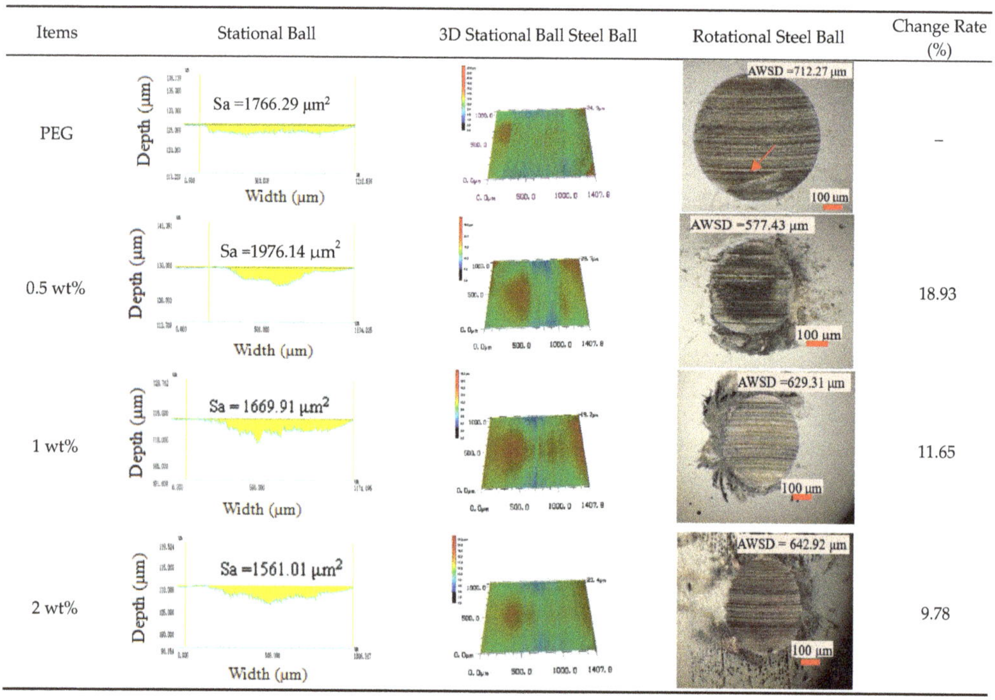

Figure 6. Cross-sectional area, three-dimensional morphology, average wear scar diameter, and change rate of wear scar area lubricated by different oil samples at 25 °C.

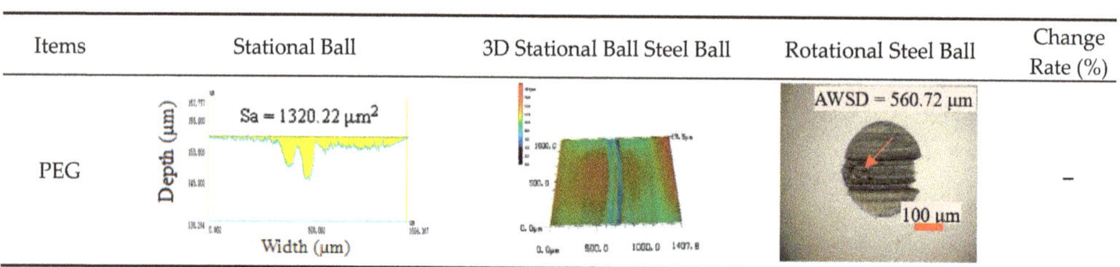

Figure 7. *Cont.*

0.5 wt%	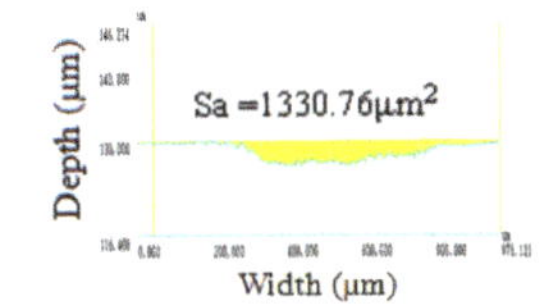	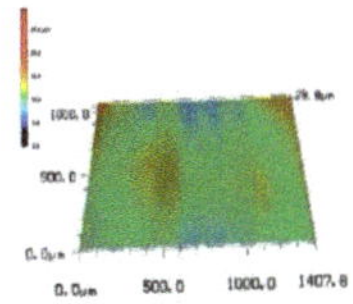	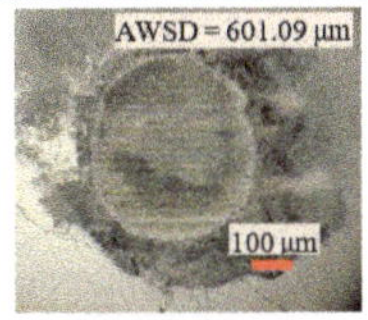	−7.2
1 wt%	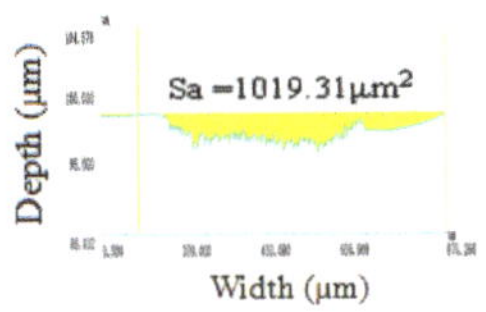	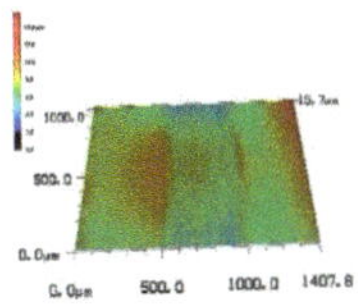	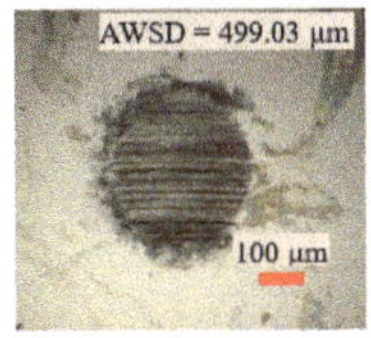	11
2 wt%	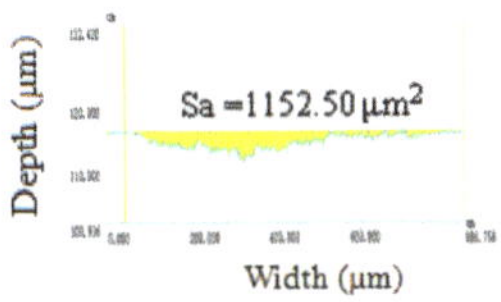	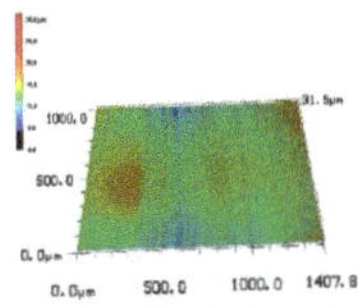	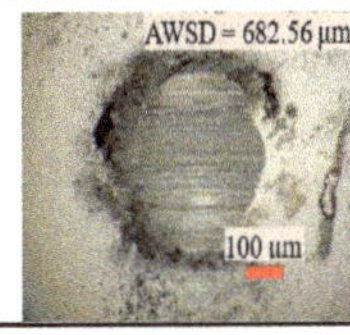	−21.73

Figure 7. Cross-sectional area, three-dimensional morphology, average wear scar diameter, and change rate of wear scar area lubricated by different oil samples at 75 °C.

Items	Stational Ball	3D Stational Ball Steel Ball	Rotational Steel Ball	Change Rate (%)
PEG	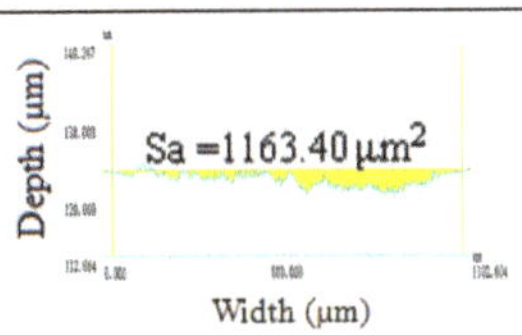	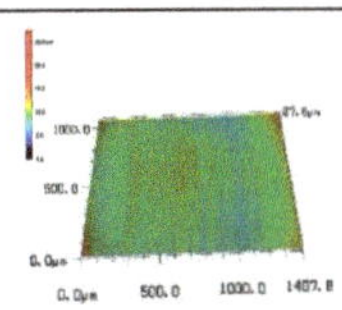	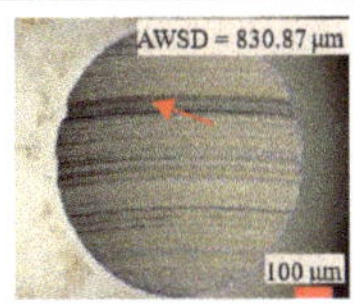	-
0.5 wt%	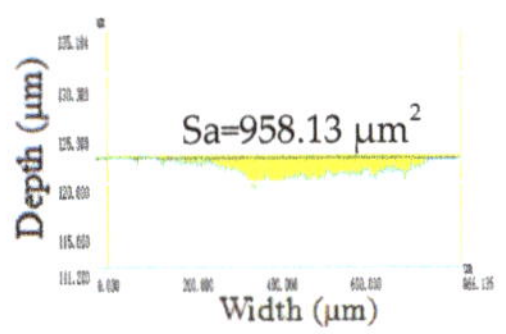	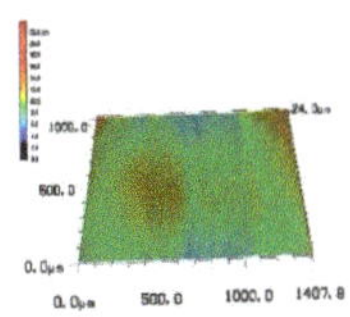	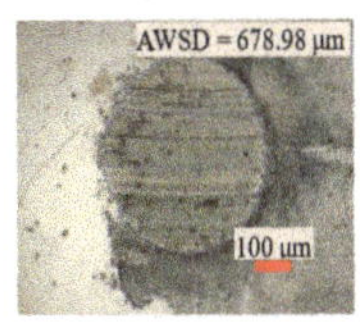	18.28
1 wt%	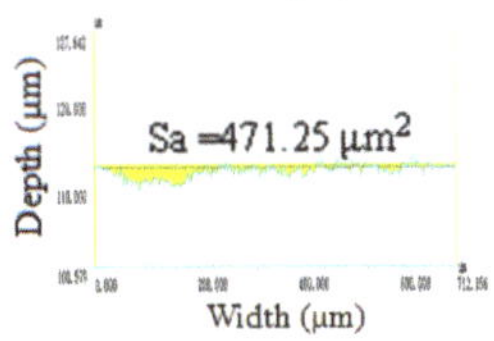	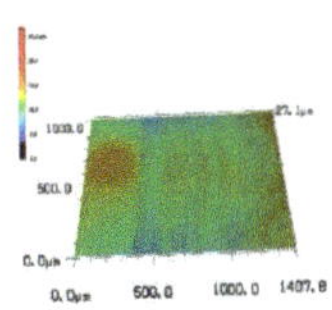	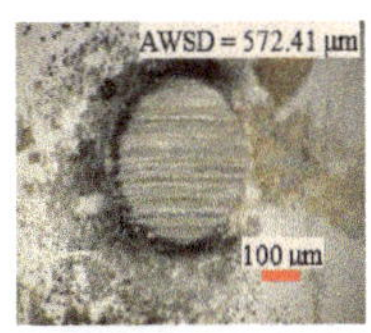	31.11
2 wt%	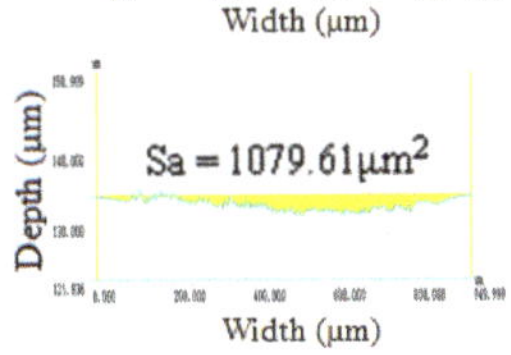	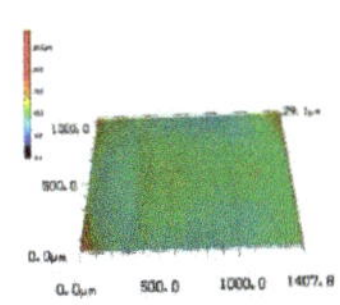	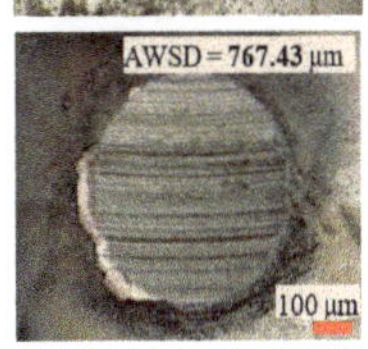	7.64

Figure 8. Cross-sectional area, three-dimensional morphology, average wear scar diameter, and change rate of wear scar area lubricated by different oil samples at 100 °C.

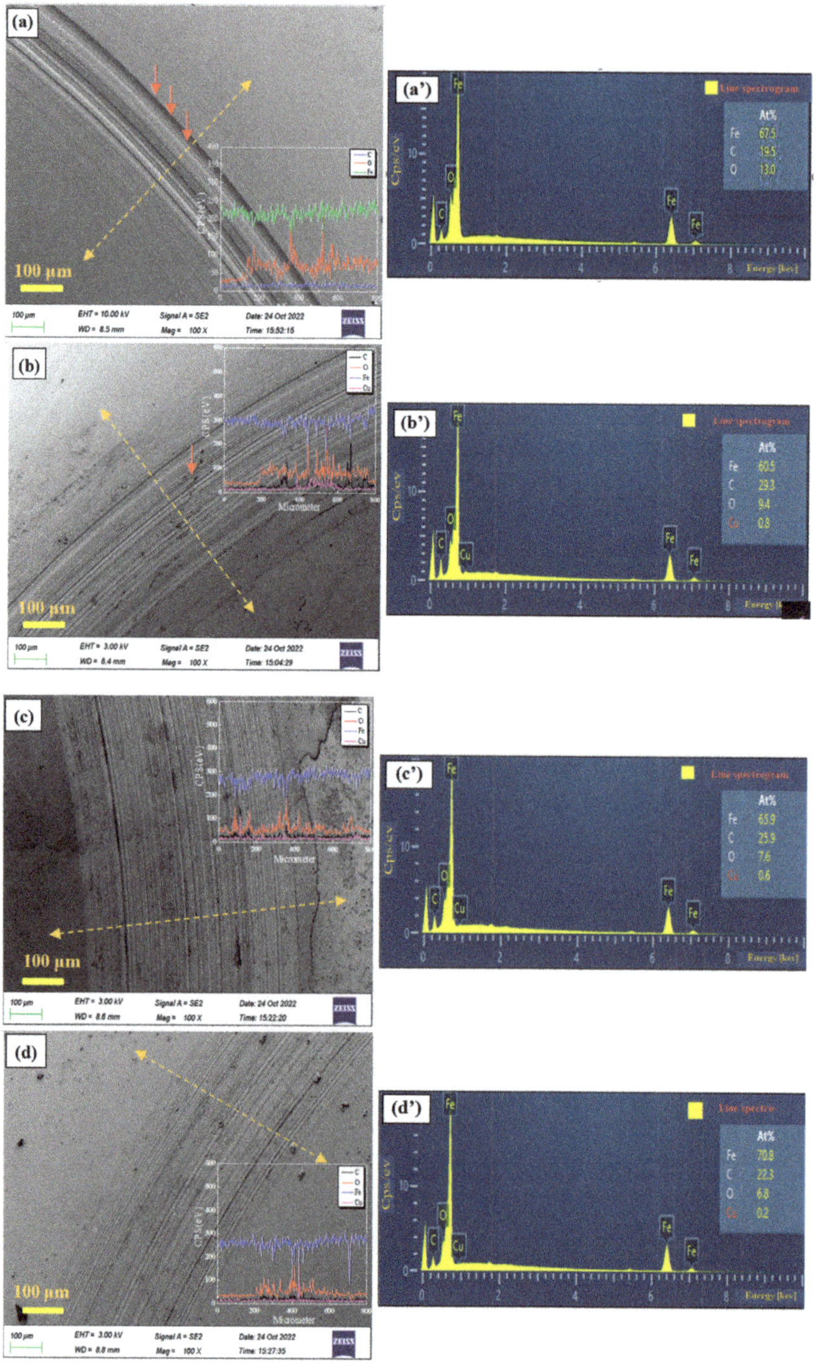

Figure 9. Surface wear scar morphology and energy spectrum analysis of steel ball lubricated by different content of copper-doped carbon, (**a**,**a′**) pure PEG, (**b**,**b′**) 0.5 wt% Cu-CQDs, (**c**,**c′**) 1.0 wt% Cu-CQDs, and (**d**,**d′**) 2.0 wt% Cu-CQDs.

3.4. Analysis of Friction and Wear Mechanism

Generally, the heat generated by friction will lead to the oxidation of lubricating oil, and the wear debris generated will also lead to further deterioration of the oil. PEG solution with different amounts of Cu-CQDs added was easy to oxidize under friction induction (Figure S3c), especially at high temperatures. The deterioration of oil products will reduce the anti-wear property of lubricating oil film. The mechanism of the lubricating performance of Cu-CQDs solution at high temperature was determined by heating the oil sample in an air blast drying oven at a certain temperature and time, and the influence of local high temperature generated at the steel ball interface during the friction process. A certain amount of Cu-CQDs solution was taken and heated to 100 °C from room temperature in an air blast drying oven. Cu-CQDs solution (2.0 wt%) was used to observe the colour change clearly. When the heating lasted for 30 min, the solution turned brick red, as shown in Figure S3b. When the solution was continuously heated or taken out of the oven for storage, the solution turned yellow again. This finding confirmed that the newly prepared 2.0 wt% Cu-CQDs produced cuprous ions when heated, but cuprous ions are not as stable as divalent copper ions in solution. The cuprous ion formed coordination compounds with water molecules in Cu-CQDs. The simulated experimental proves that the cuprous ion in solution will change into copper ion under high temperature and friction induce role.

According to the typical octahedral field and crystal field stabilization energy (CFSE) formed in the ligand field theory [35]. The energy reduction after Cu (II) coordination is relatively stable. However, the energy of Cu (I) is not stable, and the copper ion (Cu (II)) in the solution was more stable than that of copper ion (Cu (I)) [36]. When the solution was heated or taken out for standing, the unstable cuprous ion will self-disproportionate and transform into a more stable copper ion (Cu (II)). This conclusion can be used to explain the reason why the lubricity of PEG can be improved by the colour change of the oil sample during the friction process. After friction, black agglomerated CuO particles were found in the collected oil stain, indicating that the cuprous ions (Cu (I)) were oxidized to copper oxide (Cu (II)) at the interface of the friction pair during lubrication and participated in the formation of the boundary lubrication film. Figure 9 shows the EDS analysis in the wear scar area. The content of copper element increased significantly, confirming that copper oxides participated in the formation of boundary lubricating film.

The recognized lubrication mechanism of CQDs is the carbon film formed at the friction interface, and the C-σ bond passivation is related [37]. The SEM/EDS image in Figure 9 shows that the carbon content in the wear area after PEG lubrication containing Cu-CQDs (29.3 wt%) is significantly higher than that of pure PEG200 lubrication (19.5 wt%). This finding proves that Cu-CQDs additives can be rapidly deposited and embedded in the wear surface through physical absorption and electrostatic attraction [38]. With the increase in time, during the friction process, copper and trace cuprous oxide form copper derivatives on the surface of the friction pair. The continuous deposition of carbon at the interface forms a protective film (boundary friction film), thus reducing the direct contact between friction pairs and increasing the content of C and Cu. The composition of the boundary film can be illustrated by the valence states of each element in Figure 10. Figure 10a shows the general flow chart, indicating that Cu-CQDs induced by friction participate in the formation of the boundary film. Figure 10b shows the valence spectrum of carbon element (C1s) in the boundary film. The peak at 284.6 eV belongs to C-C or C-H in the film, and while the peak at 288 eV belongs to C-O. Figure 10c shows the chemical valence state of oxygen element (O1s) in the film. The peak at 529.4 eV belongs to iron oxide, while the peak at 532 eV belongs to ferric hydroxide [39]. Figure 10d shows the valence state of iron (Fe2p), confirming that iron oxides are Fe (706 eV), FeO, Fe_2O_3, and Fe_3O_4 (709–711 eV) [40,41]. Figure 10e shows the valence of copper (Cu LMM). The peak at 910 eV belongs to copper oxide, confirming that the friction induced copper-doped CQDs participate in the formation of boundary lubrication film, thus playing a good role in friction reduction and anti-wear. In addition, the low content of Cu-CQDs in PEG was dispersed uniformly which is easy to form the boundary lubrication film under friction process; however, the high content of

Cu-CQDs in PEG was easy to be agglomerated or wear debris resulted in abrasive wear. It could be proved by the aggregates in PEG after friction.

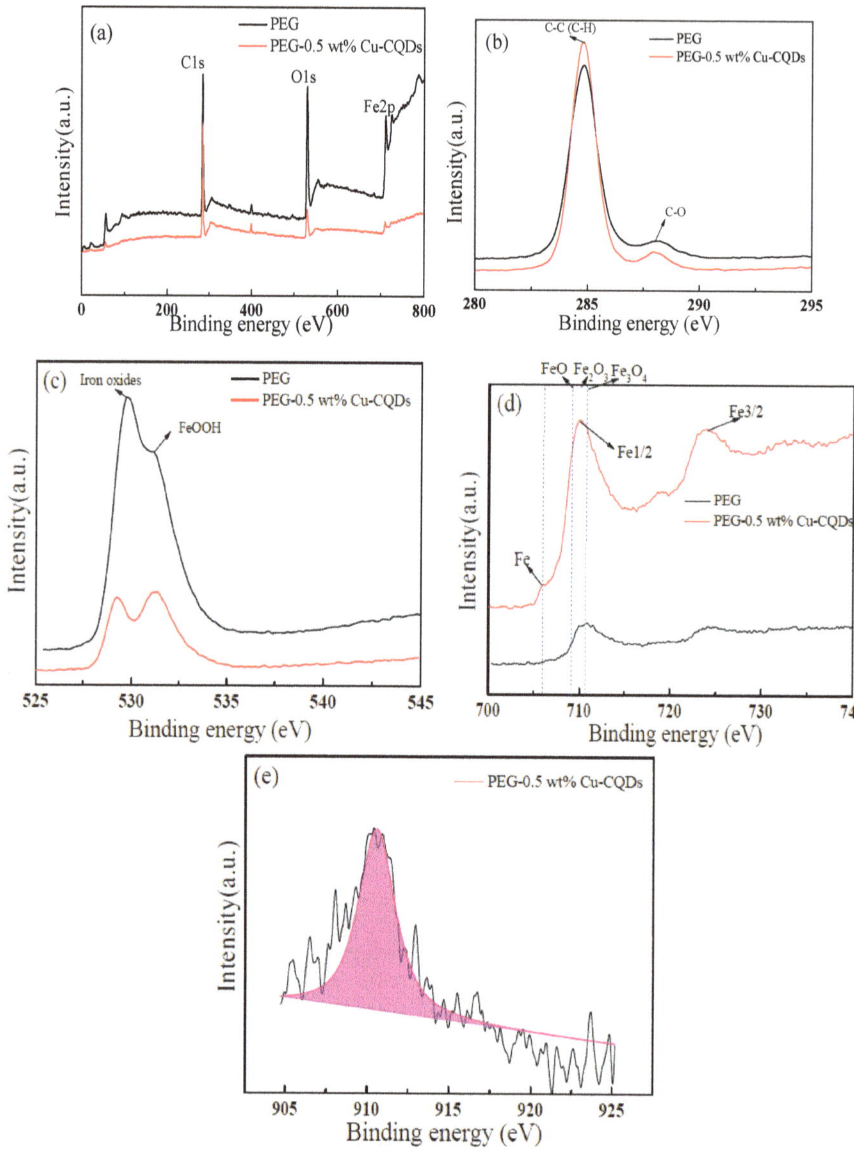

Figure 10. Valence analysis of different elements in the wear scar region of steel ball: (**a**) XPS total spectrum; (**b**) C1s; (**c**) O1s; (**d**) Fe_{2p}; and (**e**) Cu_{2p}.

4. Conclusions

High-performance lubricants for energy saving and reduced consumption were developed by preparing a novel copper-doped CQDs dispersion via hydrothermal synthesis. On the four-ball friction tester, the influence of copper-based CQD dispersion with different dosages on the lubrication performance of polyethylene glycol at different temperatures was investigated, and the following conclusions were obtained:

(1) The average particle size of the synthesized Cu-CQDs in the dispersion solution is about 8.33 nm, which shows good dispersion in PEG200.
(2) At the different test temperatures, the anti-wear and antifriction properties of PEG200 were significantly improved by adding 1.0 wt% Cu-CQDs dispersion.
(3) The friction and wear mechanism of Cu-CQDs dispersion can be attributed to the boundary lubrication film including the components of Cu-CQDs, thus reducing the friction and wear of the friction pair.

Supplementary Materials: The following supporting information can be downloaded at: https://www.mdpi.com/article/10.3390/lubricants11020086/s1, Figure S1: Fluorescence emission spectrum of Cu-CQDs; Figure S2: Infrared spectrogram analysis of CQDs and PEG (a) 1.0 wt%Cu-CQDs+PEG200; (b) pure Cu-CQDs; (c) PEG-200; Figure S3: (a) Agglomeration morphology of Cu-CQDs in PEG after friction; (b) Color of 2.0 wt% Cu-CQDs after passing through an air blast drying oven at 100 °C for 30 min; (c) Appearance color of oil stain at different temperatures of 2.0 wt% Cu-CQDs (1) 2.0 wt% Cu-CQDs, (2) 2.0 wt% Cu-CQDs at 25 °C, (3) 2.0 wt% Cu-CQDs at 75 °C, (4) 2.0 wt% Cu-CQDs at 100 °C.

Author Contributions: Methodology, S.L., K.H. and H.Z.; validation, S.L.; formal analysis, Y.C.; investigation, E.S.; data curation, X.Y. and J.W.; writing—original draft preparation, Y.X.; writing—review and editing, X.H.; supervision, E.H.; project administration, X.H. All authors have read and agreed to the published version of the manuscript.

Funding: The financial support received from the National Nature foundation of China (52075144), Anhui Province Natural Science Foundation of China (2008085ME167, 2008085QE199), Anhui University Outstanding Young Talents Programs (gxyqZD2020051) are gratefully acknowledged. The Talent Research Fund of Hefei University (21-22RC33) and Graduate Innovation and entrepreneurship program (21YCXL37).

Data Availability Statement: Not applicable.

Acknowledgments: The authors wish to express their thanks to Susheng Liu and Wuhao Jiang and Ayush Subedi for their assistance in using the SEM/EDS, HRTEM images, and 3D friction testing machine diagram and revising the language of the manuscript.

Conflicts of Interest: The authors declare no conflict of interest.

References

1. Xue, Q.; Liu, W.; Zhang, Z. Friction and wear properties of a surface-modified TiO_2 nanoparticle as an additive in liquid paraffin. *Wear* **1997**, *213*, 29–32. [CrossRef]
2. Wang, B.; Qiu, F.; Barber, G.C.; Zou, Q.; Wang, J.; Guo, S.; Yuan, Y.; Jiang, Q. Role of nano-sized materials as lubricant additives in friction and wear reduction: A review. *Wear* **2022**, *490*, 204206. [CrossRef]
3. Uflyand, I.E.; Zhinzhilo, V.A.; Burlakova, V.E. Metal-containing nanomaterials as lubricant additives: State-of-the-art and future development. *Friction* **2019**, *7*, 93–116. [CrossRef]
4. Han, K.; Zhang, Y.; Song, N. The Current Situation and Future Direction of Nanoparticles Lubricant Additives in China. *Lubricants* **2022**, *10*, 312. [CrossRef]
5. Zhang, Y. Energy efficiency techniques in machining process: A review. *Int. J. Adv. Manuf. Technol.* **2014**, *71*, 1123–1132.
6. He, Q.; Liu, H.; Ye, J. Tribological properties of nano copper as N32 lubricating oil additive. *Tribology* **2010**, *2*, 145–149. (In Chinese)
7. Qin, Y.; Wu, M.; Yang, Y.; Yang, Y.; Yang, G. Enhanced ability of halloysite nanotubes to form multilayer nanocrystalline tribofilms by thermal activation. *Tribol. Int.* **2022**, *174*, 107718. [CrossRef]
8. Jiang, H.; Hou, X.; Ma, Y.A.; Su, D.; Qian, Y.; Ahmed Ali, M.K.; Dearn, K.D. The tribological performance evaluation of steel-steel contact surface lubricated by polyalphaolefins containing surfactant-modified hybrid MoS_2/h-BN nano-additives. *Wear* **2022**, *504–505*, 204426. [CrossRef]
9. Lahouij, I.; Vacher, L.; Martin, J.; Dassenoy, F. IF-MoS_2 based lubricants: Influence of size, shape and crystal structure. *Wear* **2012**, *296*, 558–566. [CrossRef]
10. Li, B.; Xia, Y.; Wang, X. Friction and wear properties of nano Cu in polyethylene glycol solution. *Tribology* **2005**, *25*, 385–389. (In Chinese)
11. Zhu, S.; Song, Y.; Zhao, X. The photoluminescence mechanism in carbon dots (graphene quantum dots, carbon nanodots, and polymer dots): Current state and future perspective. *Nano Res.* **2015**, *8*, 355–381. [CrossRef]
12. Yang, Z.; Li, Z.; Xu, M.; Ma, Y.; Zhang, J.; Su, Y. Controllable synthesis of fluorescent carbon dots and their detection application as nanoprobes. *Nano-Micro Lett.* **2013**, *5*, 247–259. [CrossRef]

13. Kumar, V.B.; Sahu, A.K.; Rao, K.B.S. Development of Doped Carbon Quantum Dot-Based Nanomaterials for Lubricant Additive Applications. *Lubricants* **2022**, *10*, 144. [CrossRef]
14. Tang, W.; Zhang, Z.; Li, Y. Application of carbon spot in lubrication field: A review. *J. Mater. Sci.* **2021**, *56*, 12061–12092. [CrossRef]
15. Ma, W.; Gong, Z.; Gao, K. Superlubricity achieved by carbon quantum dots in ionic liquid. *Mater. Lett.* **2017**, *195*, 220–223. [CrossRef]
16. Shang, W.; Cai, T.; Zhang, Y.; Liu, D.; Sun, L.; Su, X.; Liu, S. Covalent grafting of chelated othoborate ionic liquid on carbon quantum dot towards high performance additives: Synthesis, characterization and tribological evaluation. *Tribol. Int.* **2018**, *121*, 302–309. [CrossRef]
17. Tang, J.; Chen, S.; Jia, Y.; Ma, Y.; Xie, H.; Quan, X.; Ding, Q. Carbon dots as an additive for improving performance in water-based lubricants for amorphous carbon (aC) coatings. *Carbon* **2020**, *156*, 272–281. [CrossRef]
18. Hu, Y.; Wang, Y.; Wang, C.; Ye, Y.; Zhao, H.; Li, J.; Xue, Q. One-pot pyrolysis preparation of carbon dots as eco-friendly nano-additives of water-based lubricants. *Carbon* **2019**, *152*, 511–520. [CrossRef]
19. He, J.; Sun, J.; Choi, J.; Wang, C.; Su, D. Synthesis of N-doped carbon quantum dots as lubricant additive to enhance the tribological behavior of MoS_2 nanofluid. *Friction* **2022**, *11*, 441–459. [CrossRef]
20. Ye, M.; Cai, T.; Zhao, L.; Liu, D. Covalently attached strategy to modulate surface of carbon quantum dots: Towards effectively multifunctional lubricant additives in polar and apolar base fluids. *Tribol. Int.* **2019**, *136*, 349–359. [CrossRef]
21. Tu, Z.; Hu, E.; Wang, B.; David, K.D.; Seeger, P.; Moneke, M.; Hu, X. Tribological behaviors of Ni-modified citric acid carbon quantum dot particles as a green additive in polyethylene glycol. *Friction* **2020**, *8*, 182–197. [CrossRef]
22. Huang, H.; Hu, H.; Qiao, S.; Bai, L.; Han, M.; Liu, Y.; Kang, Z. Carbon quantum dot/CuSx nanocomposites towards highly efficient lubrication and metal wear repair. *Nanoscale* **2015**, *7*, 11321–11327. [CrossRef] [PubMed]
23. Shang, W.; Cai, T.; Zhang, Y.; Liu, D.; Liu, S. One pot pyrolysis synthesis of carbon quantum dots and graphene oxide nanomaterials: All carbon hybrids as eco-environmental lubricants for low friction and remarkable wear-resistance. *Tribol. Int.* **2018**, *118*, 373–380. [CrossRef]
24. Wang, F.; Shang, L.; Zhang, G.; Wang, Z. Polyethylene glycol derived carbon quantum dots nanofluids: An excellent lubricant for diamond-like carbon film/bearing steel contact. *Friction* **2022**, *10*, 1393–1404. [CrossRef]
25. De, B.; Karak, N. A green and facile approach for the synthesis of water soluble fluorescent carbon dots from banana juice. *RSC Adv.* **2013**, *3*, 8286–8290. [CrossRef]
26. Gao, S.; Chen, Y.; Fan, H.; Wei, X.; Hu, C.; Wang, L.; Qu, L. A green one-arrow-two-hawks strategy for nitrogen-doped carbon dots as fluorescent ink and oxygen reduction electrocatalysts. *J. Mater. Chem. A* **2014**, *2*, 6320–6325. [CrossRef]
27. Hu, E.; Hu, K.; Dearn, K.D.; Hu, X.; Xu, Y.; Yu, D.; Tang, Y. Tribological performance of rice husk ceramic particles as a solid additive in liquid paraffin. *Tribol. Int.* **2016**, *103*, 139–148. [CrossRef]
28. Cai, J.; Dai, L. Study on the method for measuring the friction coefficient of lubricating oil (four ball method). *Lubr. Oil* **2001**, *16*, 5. (In Chinese)
29. Odi-Owei, S.; Roylance, B.J.; Xie, L. An experimental study of initial scuffing and recovery in sliding wear using a four-ball machine. *Wear* **1987**, *117*, 267–287. [CrossRef]
30. Ke, L.; Wang, Y. Two-dimensional sliding frictional contact of functionally graded materials. *Eur. J. Mech.-A Solids* **2007**, *26*, 171–188. [CrossRef]
31. Davey, W.; Edwards, E.D. The extreme-pressure lubricating properties of some sulphides and disulphides, in mineral oil, as assessed by the Four-Ball Machine. *Wear* **1958**, *1*, 291–304. [CrossRef]
32. Sarno, M.; Mustafa, W.A.A.; Senatore, A.; Scarpa, D. One-step "green" synthesis of dispersable carbon quantum dots/poly (methyl methacrylate) nanocomposites for tribological applications. *Tribol. Int.* **2020**, *148*, 106311. [CrossRef]
33. Wen, S.; Huang, P. *Principles of Tribology*; Tsinghua University Press: Beijing, China, 2002.
34. Sato, H.; Tokuoka, N.; Yamamoto, H.; Sasaki, M. Study on Wear Mechanism by Soot Contaminated in Engine Oil (First report: Relation between Characteristics of Used Oil and Wear). *SAE Tech. Pap.* **1999**, *1*, 3573.
35. Wang, J.; Cui, J.; Wang, X.; Qin, X. *Inorganic Chemistry*, 5th ed.; Higher Education Press: Beijing, China, 2018. (In Chinese)
36. Xia, J.; Tang, D. Preparation, characterization and catalytic performance of copper ion intercalated graphite oxide composites. *New Chem. Mater.* **2011**, *39* (Suppl. S1), 85–89. (In Chinese)
37. Mou, Z.; Wang, B.; Lu, H.; Quan, H.; Huang, Z. Branched polyelectrolyte grafted carbon dots as the high-performance friction-reducing and antiwear additives of polyethylene glycol. *Carbon* **2019**, *149*, 594–603. [CrossRef]
38. Wang, J.; Li, X.; Deng, Y.; Chen, S.; Liang, W.; Zhang, L.; Wan, Y. Carbon quantum dots doped with silver as lubricating oil additive for enhancing tribological performance at various temperatures. *Appl. Surf. Sci.* **2022**, *599*, 154029. [CrossRef]
39. Chen, Y.; Hu, E.; Zhong, H.; Wang, J.; Subedi, A.; Hu, K.; Hu, X. Characterization and tribological performances of graphene and fluorinated graphene particles in PAO. *Nanomaterials* **2021**, *11*, 2126. [CrossRef] [PubMed]
40. Hu, E.; Dearn, K.; Yang, B.; Song, R.; Xu, Y.; Hu, X. Tribofilm formation and characterization of lubricating oils with biofuel soot and inorganic fluorides. *Tribol. Int.* **2017**, *107*, 163–172. [CrossRef]
41. Hu, E.; Hu, K.; Xu, Y.; Hu, X.; Chen, D.; Cheng, Z.; Wu, D. Study on the mechanism of friction induced changes in the composition and structure of soot particles from biofuels. *Tribology* **2016**, *36*, 185–193. (In Chinese)

lubricants

Article

Different Behaviors of Friction in Open and Closed Forging Test Utilizing Palm Oil-Based Lubricants

Aiman Yahaya *, Syahrullail Samion *, Ummikalsom Abidin and Mohd Kameil Abdul Hamid

Faculty of Mechanical Engineering, Universiti Teknologi Malaysia, Skudai 81310, Johor, Malaysia
* Correspondence: wmaiman9@gmail.com (A.Y.); syahruls@mail.fkm.utm.my (S.S.)

Abstract: Increasing demand for manufactured goods in industries such as automobiles, electronics, construction, and aerospace has motivated researchers to develop sustainable manufacturing processes. Most metal-forming lubricants are not eco-friendly; they may cause substantial chemical emissions and constitute a community threat. Bio-oil lubricants are seen as possible replacements for mineral oil-based lubricants. Computational modelling of the forging process uses the finite element method to accelerate and improve design. This research intends to act as a case study and demonstrate how friction behaves differently in open-closed forging tests of different palm oil derivatives. The relationship between the different types of friction was studied using a cold forging test in conjunction with the development of a Coulomb–Tresca friction model. From the results, it can be shown that the friction behavior for the closed forging test (CFT) and the ring compression test (RCT) differs; the CFT exhibits a diversified friction adaptation, while the RCT exhibits a single friction adaptation. From both tests, palm stearin (PS) shows the lowest friction behavior where at RCT the friction is estimated at $m = 0.10/\mu = 0.05$ and the CFT has a varied friction and the average friction is estimated at $m = 0.352/\mu = 0.1626$. On the other hand, commercial metal-forming oil (CMFO) shows the highest lubrication sample in friction, where the value of friction is similar to the no lubricant sample (NA-O), which is ($m = 0.45/\mu = 0.1875$) on the RCT test and ($m = 0.424/\mu = 0.1681$) on the CFT test.

Keywords: open-closed forging; renewable material; Coulomb–Tresca friction; simulation; palm oil

1. Introduction

The use of mineral-based lubricating oil raises environmental concerns since it is known to have a high degree of toxicity and is seldom spontaneously dissolved in the environment [1]. For many manufacturing procedures, particularly those that create cold work, it is important that the die surface be lubricated. Metal-forming lubricants are used in several processes to improve tool life, metal flow, and energy efficiency by reducing friction and wear at the existing equipment contact [2]. Most commercial lubricants are based on mineral oils, which are not environmentally friendly due to their toxicity and non-biodegradability [3]. Commercial metal-forming oil also involve in open-loop system where the lubricant will end up to the surrounding.

Palm oil is one of the world's most widely used vegetable oils. It is most typically used as cooking oil, also known as palm olein. Due to its greater hydrocarbon chain length and low unsaturation, palm oil may be used as an industrial lubricant [4,5]. The length of the fatty acid chain as well as the amount of unsaturation fatty acid impact the viscosity of palm oil. Many studies have been undertaken to increase the capabilities of palm oil, but the majority of the studies has been restricted to the oil generated up to a single fractionation procedure, such as palm olein and palm stearin (PS) [5]. Many other varieties of palm oil are generated at a greater level of fractionation, such as palm mid olein (PMO). The tribological performance data for these oils is, indeed, quite limited [6].

Open die forging normally entails inserting a solid cylindrical workpiece between two flat dies and compressing it to reduce its height [7]. This process is commonly referred to as simple upsetting. In reality, the specimen takes on the form of a barrel. In most cases, barreling is generated by friction forces at the die-workpiece interfaces, which act to prevent the outward flow of material at these interfaces from occurring. The workpiece acquires the form of the die cavities (impression) as it is being agitated between the closing dies in closed-die forging, where some of the material flows radially outwards and forms flash [8]. A significant level of pressure is applied to the flash as a result of its large length to thickness ratio. As a result of these stresses, radial material flow in the flash gap is met with a lot of resistance due to the high friction caused by the material. The flash plays a crucial impact in the flow of the material in impression die forging due to the strong friction that facilitates the filling of the die cavities [9].

According to Hafis et al. [10], friction plays a significant role in the generation of stress during metal-forming. Studies have been conducted to determine the impact friction has on forming pressures and the consistency of deformed workpieces. It was suggested by Groche et al. [11] that a friction coefficient might be estimated using data from the normal metal flow that happens throughout a deformation process. There is a critical need for research into friction and its impact on tribology and open-closed forging test. Zhang et al. [12]'s friction model is often regarded via the lens of the traditional friction conceptions whenever (1) and (2)):

$$\tau = \mu p \tag{1}$$

$$\tau = mK \tag{2}$$

where,

$p = normal\ pressure$
$K = shear\ yield\ stress$
$\tau = frictional\ shear\ stress$
$\mu = Coulomb\ friction\ coefficient$
$m = Tresca\ shear\ friction.$

The majority of the earlier studies employed minerals and oils that are not suitable for human consumption in the metal-forming process. These oils and minerals include molybdenum disulfide, soap, grease, wax, rapeseed, and many more. It was found that most of this oil showed comparable results with commercial mineral-based oil [2,4,5]. Some of the lubricant has also been modified by the addition of additives, such as those investigated by Du et al., to make it meet the requirements set for commercial lubricants. [13]. It has been shown by Okokpujie et al. [8] that including nanoparticles into copra vegetable oil improves its rheological characteristics and results in less cutting force required during milling.

However, there is a limitation of study on the use of palm oil-based lubricant in metal-forming applications, where the majority of the research applications on palm oil-based were focusing mainly on experimental analysis only [6,14]. A variety of palm oil derivatives were explored and simulated using the finite element method (FEM) to assess as metal-forming lubricant potential. This was done to ensure that the researchers could observe the frictional behavior of each sample test. This study aims to fill a research gap by comparing friction behavior under various types of forging processes using the best palm oil derivatives based on their physicochemical properties by modelling finite element method of sample lubricant with different applications of the forging process (Open and closed forging).

2. Methodology

2.1. Physicochemical Properties of Sample Lubricants

Table 1 summarizes the physical parameters of all six kinds of palm oil-based lubricants. The six types of palm oil include refined, bleached and deodorized palm oil (PO), palm fatty acid distillate (PFAD), palm olein (PL), palm stearin (PS), palm kernel oil (PKO), and palm mid-olein (PMO).

Table 1. Physical properties of palm oil [15].

Lubricant	PKO	PO	PS	PL	PMO	PFAD	CMFO	Test Method
Density (kg/m^3) @ 25 °C D1298-85(90)	0.887	0.880	0.870	0.890	0.895	0.873	0.900	ASTM
Kinematic Viscosity (mm^2/s) @ 25 °C	45.77	215.47	48.29	46.74	54.6	120.14	146.24	ASTM D445-94
Kinematic Viscosity (mm^2/s) @ 40 °C	35.36	189.40	38.01	35.00	50.6	96.35	107.71	ASTM D445-94
Kinematic Viscosity (mm^2/s) @ 100 °C	11.24	18.20	8.56	14.4	13.2	8.9	11.2	ASTM D445-94
Viscosity index, (VI)	329	106.00	213	426	272	48	88	ASTM D2270
Pour point (°C)	21.0	34.0	37.3	9.0	15.0	35	−24.0	ASTM D9793
Melting point (°C)	27.5	48.50	44.0	21.5	18.50	38.0	-	-
Flash Point (°C)	205	205	315	318	324	135	236	-
Cloud Point (°C)	-	-	-	9.5	18.5	-	-	-
Iodine value (WIJS)	17.8	17.8	27.8	56.15	48.23	24.80	-	-
Free fatty acid (%)	0.06	0.06	0.05	0.07	0.15	86.4	-	ASTM D664
Peroxide value (PV)	0.84	0.85	1.04	0.95	0.92	1.58	-	-

One measure of a fluid's resistance to movement through its environment is its viscosity. Utilizing oil with a higher viscosity resulted in a thicker oil coating; however, this increased flow resistance and reduced the ability to protect the contact surface [6,15]. It is possible that low-viscosity oils may help in the boundary lubrication regime, but in the sliding system, they'll be rapidly worn away by the contact surfaces. Lubricant oil with a high viscosity index (VI) is generally considered as a good lubricant [14,15]. Higher VI meant that the viscosity of the oil did not vary much when the temperature is raised. That is, the oil's viscosity profile suffered fewer phase shifts, resulting in greater heat stability. In terms of viscosity index, PKO had the greatest value of 329 and commercial metal-forming oil had the lowest value of 88. The slip melting point is another essential characteristic that indicates lubricant oil fluidity. In the palm oil sector, the slip melting point was the temperature at which all crystal nuclei were melted, which occurred most often in the stearin fraction. The degree of unsaturation of fatty acid molecules is determined by the iodine value (IV) [16]. It is an essential parameter since it reveals the fats and oils' properties. IV is highly correlated with the cloud point and the fatty acid makeup of the oil.

The top three varieties of palm oil-based lubricant were chosen using the scoring system shown in Table 2. The score is determined by the following rank (1 is good to 6 is poor) and is based on the reasoning provided in Table 1. The oil with the lowest score was rated first, followed by the others.

Table 2. Evaluation score of palms oil-based lubricant.

Lubricant	PKO	PO	PL	PS	PMO	PFAD
Density (kg/m^3) @ 25 °C	3	4	5	1	6	2
Viscosity index, (VI)	2	5	1	4	3	6
Pour point (°C)	3	6	5	1	4	2
Melting point (°C)	2	5	6	1	3	4
Flash Point (°C)	5	4	3	2	1	6
Iodine value (WIJS)	6	2	1	4	3	5
Free fatty acid	1	4	6	2	3	5
Peroxide value	1	2	4	5	3	6
Total score	23	32	31	20	26	36

According to Table 2, PS leads the pack with a total score of 20 and is followed by PKO at 23. The cloud point feature was not taken into account in this score assessment since it cannot be identified in certain oils. It may be concluded that PS, PKO, and PMO have shown acceptable basic lubricant oil characteristics and have been chosen as a palm oil-based lubricant in the metal-forming processes.

Gas-liquid chromatography (GLC) was used to determine the chemical characteristics in terms of fatty acid composition (FAC) as shown in Table 3. The sample was separated into its constituents depending on their affinity for the liquid and mobile phases [16]. According to Table 3, PKO has the largest amount of saturated fatty acid, coming in at 82.3%, while PMO has the least amount, falling in at 53.5%. PS and PMO have a pattern that is comparable, in that they have a high concentration of palmitic and oleic acid, in contrast to PKO, which has a high concentration of lauric acid. This can be seen by looking at each composition of the fatty acid. Figure 1 shows the experimental set-up for both testing (open and closed forging).

Table 3. Composition of fatty acids in palm oil used as a lubricant [2].

FAC (% by Gas Chromatography)	PS	PKO	PMO
Caprylic acid C8:0	-	3.6	-
Capric acid C10:0	-	3.5	-
Lauric acid C12:0	0.16	47.8	0.5
Myristic acid C14:0	1.16	16.3	1.1
Palmitic acid C16:0	54.31	8.5	45.0
Stearic acid C18:0	4.71	2.6	6.4
Oleic acid C18:1	32.31	15.3	37.3
Linoleic acid C18:2	6.68	2.4	8.8
Linolenic acid C18:3	0.3	-	0.2
Arachidic acid C20:0	0.37	-	0.5
Eicosenoic acid C20:1	-	-	0.1
Saturated fatty acid	60.71	82.3	53.5
Mono-unsaturated fatty acid	32.31	15.3	37.4
Poly-unsaturated fatty acid	6.98	2.4	9
IV	33	17.8	54

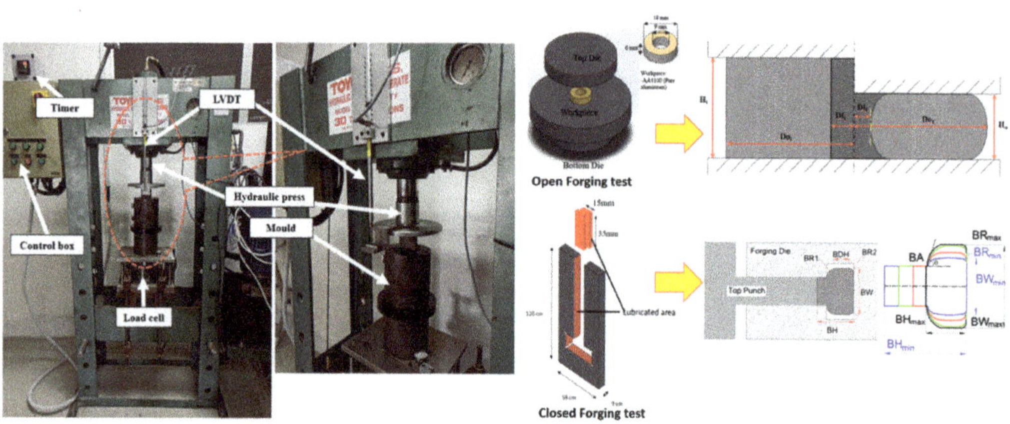

Figure 1. The open-closed forging test set up [15,17].

2.2. Metal-Forming Test

In order to evaluate the frictional behavior with various derivatives of palm oil-based lubricants, two main tests were conducted, which were open forging and closed forging test. A hydraulic press machine, similar to the one seen in Figure 1, was used to conduct an analysis on each and every test sample. The experimental materials and conditions are summarized in Table 4.

Table 4. Materials and conditions for experiments [14,17].

Properties	Open Forging Test	Closed Forging Test
Workpiece	Pure aluminum (A1100)	
Workpiece Hardness (Hv)	Before annealing = 134.8 After annealing = 52.6	
After annealing = 52.6		
Tooling material	SKD-11	
Workpiece size (mm)	18:9:6	35 × 15 × 4.5
Reduction in height	~10%, ~20%, ~30%, ~40% and ~50%	
Lubricant quantity (mg)	~5 mg	~10 mg
Temp (°C)	24–27 (Room temperature)	
Compression speed (mm/s)	1	

2.3. Finite Element Method

The DEFORM-3D optimization routines employ three types of variables while working toward an optimum design: design variable (independent variable), state variable (dependent variable) and objective function. The height is defined as a design variable with an initial value of 35 mm and the billet width as a function of the billet height. The state variable are quantities that are used to define a constraint for the design and also known as a dependent variable with typical response quantities such as stresses and displacement [18]. Objective function in closed forging die is projected to the complete die filling without defects. The flow of the A1100 in the mold is analyzed throughout this study, and the aim of the simulation is to identify the maximum possible height over which the material may flow. In the DEFORM-3D parametric design language, they are provided as scalar parameters [19]. In an optimization study, the independent variables are defined, such as tool and die geometry, lubricant, friction, and amount of deformation. Variables in the design are represented by their vectors, which are:

$$x = [x_1\ x_2\ x_3\ x_4\ \ldots\ x_n\ ] \tag{3}$$

Each design variable has n associated constraints, or maximum and minimum values;

$$\underline{x_i} \leq x_i \leq \overline{x_i} \qquad (i = 1,2,3,\ldots,\ n) \tag{4}$$

where:

n = number of design variable.
x_i = Design variable
$\underline{x_i}$ = Lower limit
$\overline{x_i}$ = Upper limit

The limits placed on the design variables are known as side constraints, and they set the bounds that define the so-called viable design space.

Now, minimize

$$f = f(x) \tag{5}$$

Subject to

$$g_i(x) \leq \overline{g_i} \qquad (i = 1,2,3,\ldots,\ m_1) \tag{6}$$

$$\underline{h_i} \leq h_i(x) \qquad (i = 1,2,3,\ldots,\ m_2) \tag{7}$$

$$\underline{w_i} \leq w_i(x) \leq \overline{w_i} \qquad (i = 1,2,3,\ldots,\ m_3) \tag{8}$$

where:

f = objective function

g_i, h_i, w_i = state variables containing the design, where under- and over-bars denote minimum and maximum values (input as max and min on optimization variable commands). m_1, m_2, m_3 are the total number of constraints placed on state variables by their respective ranges of allowable values. Due to their correlation with the vector (x) of design variable, state variables are often called "dependent variables". The purpose of the constrained minimization problem represented by Equations (3) to (5) is to minimize the objective function $f(x)$, subject to the restrictions given by Equations (6) to (8).

Correlation between real stress–strain data acquired from pure aluminum A1100 workpiece using the Instron 5982 universal testing equipment is used to describe the material. The results of the tensile tests performed on the three samples given for this study are shown in Figure 2. Table 5 shows the parameters that are used in FEA of both openforging test (RCT) and closed forging test (CFT). As suggested by Li et al. [20], the following may be used as a first estimate for the plastic region curve of stress–strain:

Table 5. Parameters used in FEA of forging test.

ANALYSIS OPTIONS	FEA Program	Scientific Forming Technologies Corporation (SFTC)	
	Compiler for User Subroutines	DEFORM-3D Ver-10.2	
	Material Plasticity Method	Elastic-Plastic formation	
	Die Material Type	Rigid	
	Punch Velocity	1 mm/s	
	Number of steps	RCT	CFT
		50	400
	Iteration Method	Newton-Raphson	
	Remeshing	Global remeshing, overlay quad type, depends on element distortion,	
	Number of meshing	RCT	CFT
		55000	45000
	Deformation	Active in FEM + meshing	
	Temperature	20 °C	
CONTACT	Inter-Object Data Definition	Coulomb-Tresca friction model	
	Relative Sliding Velocity	Default (=0)	
	Tolerance	0.0256	
MATERIAL		Model	
	Young's Modulus	A1100 = 68.900 GPa	
	Poisson's Ratio	A1100 = 0.33	

It can be observed in Figure 2 that stress data are only accessible up to a value of $\varepsilon_{true} = 0.032$ in this case. However, for this extension, the smallest size as well as the chord length at the neck are required. Because tensile stress causes metals to lose their formability, upsetting is used to increase the strain values of the metals under test. If the flow curve is simply to be calculated for low stresses, the tensile test will offer sufficient information for this purpose. If the flow curve calculated for low strains can be extended to greater strains, this test is even effective in certain conditions. The stress–strain curve in the plastic region can be approximated by:

$$\sigma = H\varepsilon_p^n \tag{9}$$

$$\varepsilon_p = \left(\frac{\sigma}{H}\right)^{1/n} \tag{10}$$

where

ε_p = plastic strain

H = strength coefficient
n = unitless pressure hardening exponent

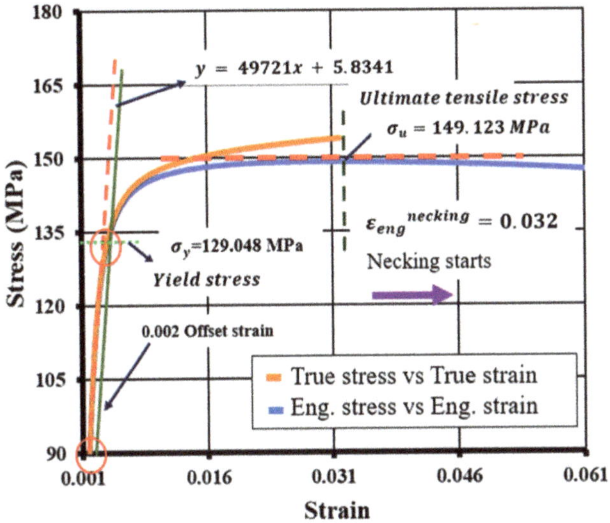

Figure 2. The tensile testing resulted in a stress–strain curve for AA1100 [15].

It was proposed by Ramberg and Osgood that this connection may be utilized to anticipate the shape of the stress–strain curve for any given material. Tensile and yield strengths of this plastic may be determined using the following equations [21];

$$\sigma = 140.82244(\varepsilon)^{0.01405} \tag{11}$$

where
σ = true stress
ε = true strain.

The design of the rig was based on the assumption that all of its components were solid bodies, which was made possible by the use of a stiff surface analysis. The aluminum alloy utilized in this study is one that, as noted by Szala et al. [22], has generally been a strain-hardening and cold-forming material.

3. Results and Discussion

3.1. Open Forging Test

The effects of several palm oil-based lubricants on load variation throughout the forging process were investigated, and the results were compared to those obtained without lubrication (NA-O) and a CMFO lubricant used as a benchmark. A direct relationship between force and die stroke was discovered in experiments, as shown in Figure 3. The graph displays the initial deformation as well as the after deformation, with the early deformation beginning at 0 kN and ending at 3 mm (50%) compression of the die stroke with varying load values for each sample.

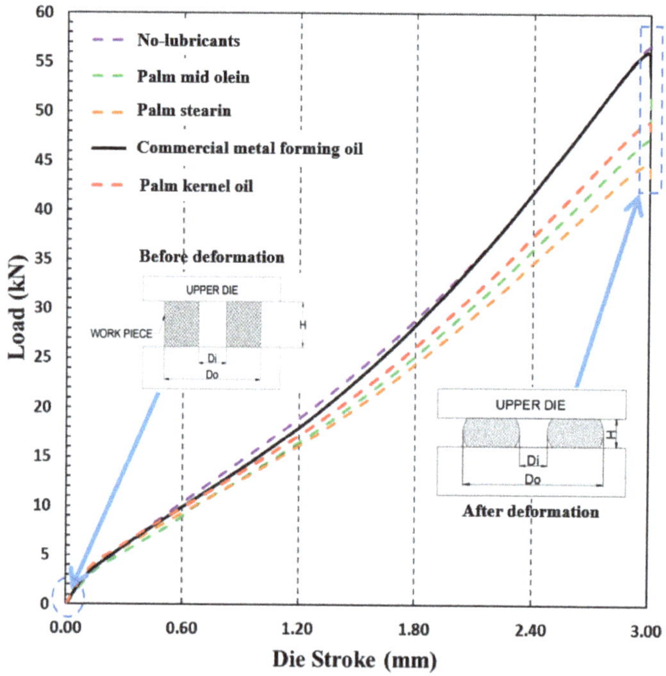

Figure 3. Testing the forging force of a NA-O, PKO, PS, PMO, and CMFO.

From the result obtained, the compression load is directly proportional to the die stroke. The absence of lubrication in the workpiece and the interaction between the metals have both indicated a growing load, and it is clearly evident that the NA-O sample had the largest degree of compression during the whole of the test where the max load is around 56.71 kN. According to Tiong et al. [23], the presence of lubricant normally contributes to the reduction of friction, which consumes less compression load; however, the presence of CMFO in ring comparison test shows that the load somewhat is comparable to the NA-O sample, where the load is lower only at 10% to 30% die stroke with max compression load of 55.99 kN. Compared to the benchmark lubricant (CMFO), all of the samples of the palm oil-based lubricant show a considerable reduction in compression load, indicating that it has the potential to be used as a lubricant in metal-forming process. PS, on the other hand, demonstrates to be the most desirable lubricant in RCT tests with the max compression load at 44.85 kN, where the compression load exhibited the lowest load when compared to other palm oil-based lubricants. PKO is a well-known palm oil-based lubricant that has been tested as an engine oil lubricant and in some cases, such as the one proposed by Aiman et al. [17], the lubricant performed better than the mineral oil. PMO is also one among the products of palm oil, but it has received less attention from researchers than other palm oil products, particularly in the metal-forming process. This product has shown some intriguing findings, which include a reduced compression load (48.56 kN) in comparison to PKO (49.08 kN) and CMFO sample in RCT test.

A lubricant is acceptable for use on the contact surface during extrusion if it keeps the extrusion load within an acceptable range, as stated by Wang et al. [24]. This is one of the criteria for determining whether or not a lubricant is suitable for this application. In these situations, the reduced compression load of PS, PKO, and PMO might be related to the material's physical behavior, in which it exists in a semi-solid form at ambient temperature and completely liquidizes when heated to 40 °C [25]. As highlighted by Maleque et al. [26], palm oil's high concentration of free fatty acids (FFA) contributes to the creation of a thin lubricating coating between the taper die and the work material. According to the data shown in Table 3, palm oil seems to include oleic acid, which may aid in reducing the amount of drag encountered by moving objects.

The strength of a fatty acid chain is determined by how the densely packed alkyl chain reacts to the accumulated short range Van der Waals interactions between neighboring methyl groups [27]. A greater affinity on the metal surface was achieved with a higher closed packed density. The ninth and tenth carbons of the fatty acid's chain are double-bonded, making it unsaturated. Confirmed by Campen et al. [28], this double bond causes the unsaturated fatty acid oleic acid to adopt a cis-configuration, making it difficult for the molecules to adopt a linear shape. Unsaturated fatty acids are therefore far less efficient in forming tightly packed monolayer soap films. Less compact packing density causes fatty acid chain molecules to have less attraction for metal surfaces. Meanwhile, the molecules of palmitic and stearic acid, which are saturated fatty acids, have a great capacity to pack tightly and effectively on metal surfaces (Wood et al., 2016). The metal surface was protected by PS (C16:0 = 54.31% and C18:0 = 4.71%) and PMO (C16:0 = 45.0% and C18:0 = 6.4%) with a high amount of saturated fatty acids, which also showed excellent molecular packing and offered a decreased compression load compared to PKO as discussed by Zulhanafi et al. [6].

Calibration Curve and Optimization Process

As a means toward a more complete understanding of frictional behavior, it is crucial to compare the experimental result with the findings of the FEM. In a mathematical optimization problem, the objective function is the real-valued function whose value is to be reduced or maximized relative to the set of feasible solutions. Figure 4 depicts the comparison of exploratory results with the contact calibration curves, which was used to determine the TSF and CFC. For the purpose of identifying optimal interaction parameters, Zhang et al. [29] claim that there is no widely accepted method for matching experimental data with the optimum calibration curve. As previously indicated [13], Equation (12) was utilized as a benchmark to establish the correlation of friction between Tresca shear friction (TSF) and Coulomb friction coefficient (CFC).

$$\frac{|d_\mu - d_m|}{d_0} < e \tag{12}$$

where

d_μ = inner diameter of the CFC
d_m = inner diameter of the TSF
d_0 = inner diameter before deformation
e = positive smaller of the two inner diameters.

A lower positive value e is obtained by comparing the inner diameters of the specimens. Because of the precision of the finite element model, Zhang et al. [13] noted that the inner diameter of the specimen remains almost constant for $e < 0.005$ despite a little fluctuation in friction. As a result, $e = 0.005$ may be a good figure to use to determine whether the narrowing of the metal ring's inner width agrees with the appropriate calibration curve. Using Equation (12), we can define Equation (13) to evaluate the friction parameters, and we know that $e = 0.005$ or 0.5% is the smallest positive value that best describes the situation.

$$\frac{|D_{Exp} - D_{\mu,m}|}{D_0} = |\delta D_{Exp} - \delta D_{\mu,m}| < e \tag{13}$$

where

D_{Exp} = inner diameter of experimental data
δD_{Exp} = change in inner diameter of experimental data.
$D_{\mu,m}$ = inner diameter of each friction representation

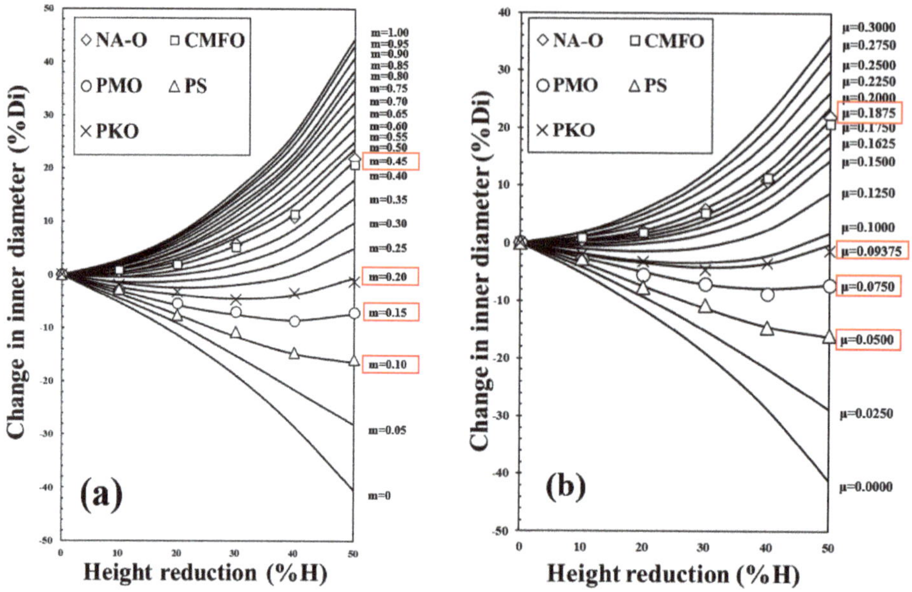

Figure 4. Validation of experimental result of change in inner diameter for each height reductio (**a**) Tresca shear friction and (**b**) Coulomb friction coefficient.

The internal diameter changes brought about by a minor shift in μ or m for a moderate reduction in height are negligible. The discrepancies between the test findings and the fitted calibration curves were determined and analyzed using Equation (13).

Figure 4 depicts the calibration curves for each of the samples after the e values were determined for each one and compared to the FEM. The line in this figure represents the simulation data, while the point in the figure is the actual experimental result. The data reveal that the friction is best matched at $m = 0.45$ and $\mu = 0.1875$ for the no lubricant sample (NA-O), indicating that this sample has the maximum frictional value. When compared to the deformation shown in the NA-O sample, the CMFO result shows a similar trend. The PS, on the other hand, had the least amount of friction at $m = 0.1$ and $\mu = 0.05$, as shown by the findings from the sample that was based on palm oil. When compared to the PS, the PMO exhibits a somewhat greater amount of friction where the values for friction are $m = 0.15$ and $\mu = 0.075$. PKO demonstrates the greatest amount of friction in palm oil based on the values of $m = 0.20$ and $\mu = 0.09375$.

Data from the cold work compression test with the lubricant sample are compared with the force applied in the FEM to validate the friction coefficient. The steady-state effects of force owing to displacement were reported in the experimental investigation at a variety of stroke locations and compared with the FEM simulations with varying shear friction factors.

Figures 5 and 6 show the result of comparing the TSF and CFC calibration curves with the validation of compression load results; this shows that all of the sample tests have a strong connection with a single friction behavior. According to the findings (Figure 5a,b), there is a clear connection between the friction of the NA-O sample and the value of $m = 0.45/\mu = 0.1875$, which is the same as the value for the CMFO sample. From Figure 5e, PS showed the lowest fiction and has a high connection to the $m = 0.10/\mu = 0.05$, whereas the friction behaves $m = 0.15/\mu = 0.075$ for PMO (Figure 5c). PKO exhibits the lowest amount of friction in comparison to the palm oil-based lubricant, in which the compression load is precisely equivalent to the $m = 0.20/\mu = 0.09375$ expression.

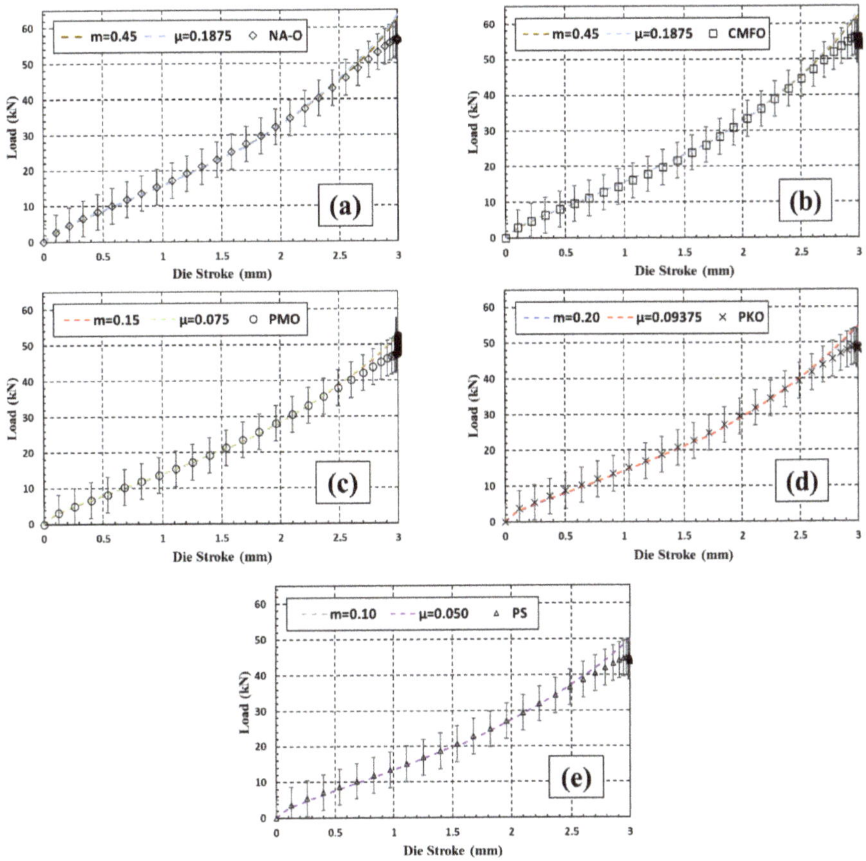

Figure 5. Force against height of deformation: FEM modelling and experimental data for (**a**) NA-O, (**b**) CMFO, (**c**) PMO, (**d**) PKO, and (**e**) PS [17].

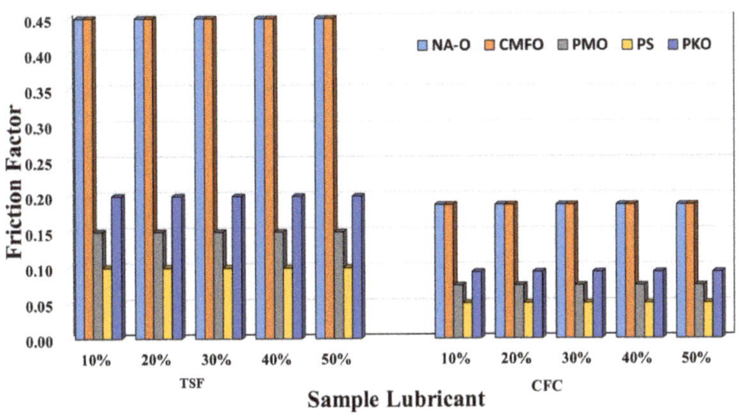

Figure 6. Summary of friction behavior for each component of the sample test following FEM optimization for open forging test.

The data that were collected throughout the process of calibration and optimization may be used to determine the efficiency of each sample. Equation (14) has been used to estimate the effectiveness of the lubricant in terms of the efficiency of friction lubrication percentage (m = 0, μ = 0 as a 100% efficacy that is denoted as $m\mu_0$). This comparison is made in relation to the situation where there is no lubricant present (NA-O) (Zhang et al., 2019). Where $m\mu_{NA-O}$ = friction of non-lubricated sample and $m\mu_{sample}$ is the friction that occurs in each lubricated sample.

$$\left| \frac{(m\mu_{NA-O} - m\mu_0) - \left(m\mu_{NA-O} - m\mu_{sample}\right)}{(m\mu_{NA-O} - m\mu_0)} \right| \times 100\% = efficiency\% \tag{14}$$

According to the results of the calculation of efficiency provided by the CMFO, there is no improvement where the friction is maintained at the same level as in the NA-O sample. The PS, on the other hand, has the highest efficiency, which decreased the friction by around 77.8%, while the improvement for the PMO was approximately 66.7%. PKO, on the other hand, demonstrates a worse performance in comparison to PS and PMO, which is somewhere about 55.6%.

According to Caminaga et al. [30], a lubricant may be acceptable for usage in the deformation zone during the extrusion phase if the lubricant could lower the extrusion stress to a decent degree. It has been demonstrated that as metal-to-metal contact increases, the mechanism requires more energy to slide the material and raise the compression load due to higher friction. Besides that, NA-O and CMFO shows that the presence of lubricant does not contribute to an improvement in the friction behavior in these areas.

As per Tiong et al. [23], PS is semi-solid at ambient temperature, would entirely liquidate till it crosses 40 °C, and may display sluggish mobility during the forging process where PMO also conduct a similar characteristic. These physical conditions reduce friction and compression force during transformation. The amount of work piece that came into contact with the die throughout the process was small. According to the findings of a previous research, one of the benefits of having a higher viscosity is that it may reduce the amount of damage and wear that is caused to surfaces by retaining the layers that are generated between the rubbing surfaces [31]. The shift from a mixed to a boundary lubrication regime is facilitated by a lower viscosity, which results in the surface being subjected to a greater coefficient of friction and increased wear [16]. This behavior may account for the fact that PKO, as shown in Table 1, has higher friction due to its lower viscosity compared to the other palm oil-based lubricant [31].

Moreover, FFA-penetrating palm oil-based lubricant provides a thin fluid film-lubricating layer between the work material and taper die, as described by Maleque et al. [26]. The oleic acid content of palm oil is shown in Table 2, which may decrease relative motion fiction. Since glycerol normally binds to fatty acid to generate the thin film between both the surfaces under rubbing action, higher rates of glycerol and fatty acid contribute through a thicker layer, decreasing friction, wear, and load of extrusion, as stated by Yingying et al. [32]. A longer carbon chain in the FFA composition indicates a higher concentration of FFA, as shown by the aforementioned study, which demonstrates that the PS has potential as a metal-forming lubricant.

3.2. *Closed Forging Test*

The effects of several types of palm oil-based lubricant, as well as no lubrication and commercial metal-forming lubricant, on the load variation of the forging process were investigated. Figure 7 demonstrates the experimentally established connection between punch load and die stroke. In the first stage, a comparison of the sample lubricants reveals that all of them follow a fairly identical trajectory, in which the load abruptly escalates. A unique pattern emerges in the second phase of the lubricant's influence. By measuring how much friction is reduced during the compression test, researchers can now estimate how well the lubricant sample performs. When compared to alternative lubricants, palm stearin (PS) has shown superior lubrication efficiency, with the lowest compression load appearing in the third stage and remaining constant all the way through to the final product production. A study by Tiong et al. [23], found that at ambient temperature PS has the consistency of a semi-solid but may become a fluid at temperatures of 40 °C or beyond. This quality may aid lubricants in decreasing the amount of metal-to-metal contact that takes place by decreasing the amount of friction between metallic surfaces. Comparatively, the compression load of CMFO is much higher than that of other palm oil-based lubricants, such as PKO and PMO. The iodine and moisture found in palm oil have been demonstrated to provide anti-friction properties between two moving metal substrates [33]. Reduced friction and compression force during the transitional stage suggest that as little metal as possible was brought into contact with the die and the workpiece.

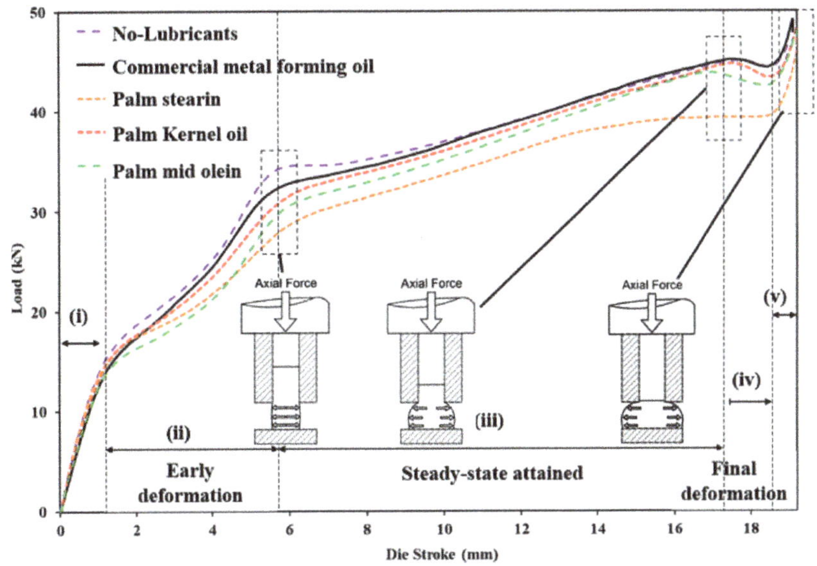

Figure 7. Compression load of NA-O, CMFO, PS, PKO, and PMO at certain die stroke.

Table 1 demonstrates that oleic acid, which is assumed to be present in palm oil, could assist in lowering relative motion blur. Moreover, free fatty acids (FFA) penetrating palm oil aid in forming a thin lubrication coating between the taper die and the workpiece material, as indicated by Maleque et al. [26]. Glycerol normally binds to fatty acid to produce a thin film between the two surfaces when rubbing happens; increased quantities of glycerol and fatty acid contribute through a thicker layer, which lowers friction, wear, and compression stress, as stated by Yingying et al. [32]. The aforementioned analysis reveals that the PS is appropriate since the FFA content has a much longer carbon chain than the PMO and PKO.

3.3. Optimization and Analysis of FEM

The compressed sample's billet width and billet height were measured, along with a few other design factors, so that test results could be compared to those obtained from the simulation. Furthermore, as was previously said, there is no universally accepted method for determining how to best correlate experimental data with the appropriate calibration curve, which is necessary for determining interaction conditions [29]. As stated by Zhang et al., [12] Equations (15)–(17) was utilized as a benchmark in order to assess the form of the relationship that exists between TSF and CFC.

$$\frac{\left|BW_{exp} - BW_{\mu,m}\right|}{BW_0} = \left|\delta BW_{Exp} - \delta dBW_{\mu,m}\right| < e \tag{15}$$

$$\%BW = \frac{BW_o - BW_{exp,m,\mu}}{BW_o} \times 100\% \quad (16)$$

$$\%BH = \frac{BH_o - BH_{exp,m,\mu}}{BH_o} \times 100\% \quad (17)$$

where

BW_{exp} = experimental width
$BW_{\mu,m}$ = width for the specimen of TSF or CFC model
BW_0 = width before deformation
e = smaller positive number from the specimen comparison

With the use of Equation (15), the gaps that were found between the results of the experiments and the fitted calibration curves were computed and illustrated in Figure 8. The examination concluded that the experimental results were in close agreement with one calibration curve, with just a few minor discrepancies amounting to less than 0.5% of the total.

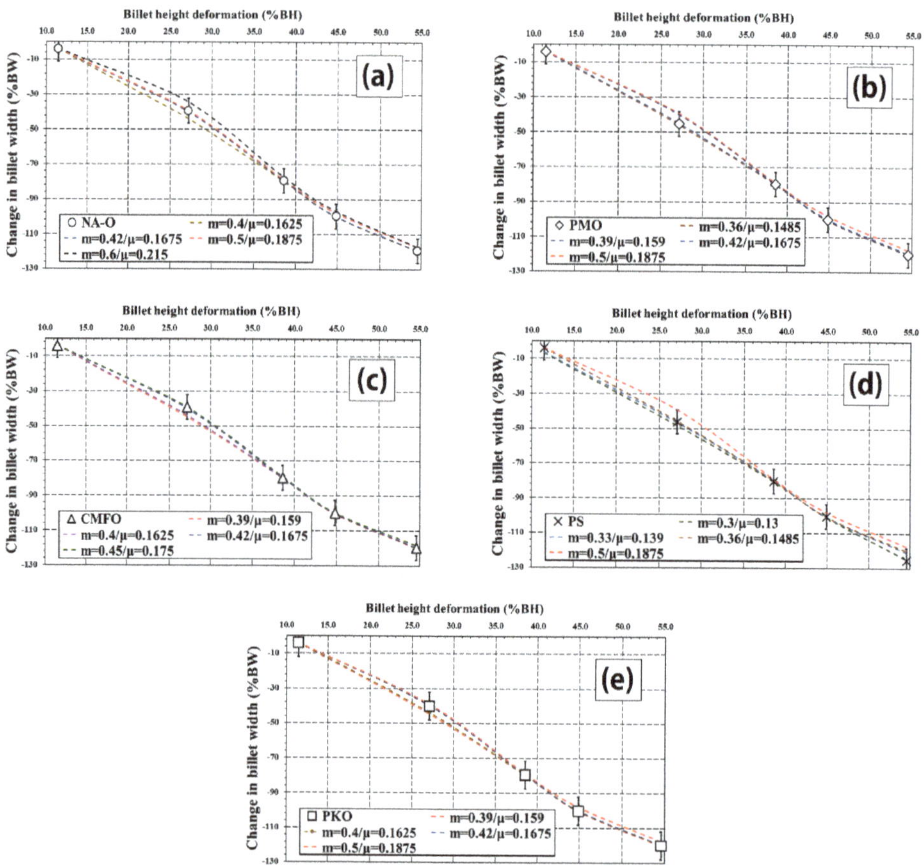

Figure 8. Calibration graph of closed forging test for (**a**) NA-O, (**b**) PMO, (**c**) CMFO, (**d**) PS, and (**e**) PKO.

Into a more precise calculation of the lubricant sample's friction coefficient, the FEM is used to compare the collected data to the applied force during the cold work compression test. Figure 8 shows the results of an experiment measuring the effects of displacement on force at steady state, while Figure 9 displays the results of a comparison between the

FEM simulations and the experiments conducted under different CFC and TSF. Given the current arbitrary Lagrangian–Eulerian (ALE) configuration, the FEM was used to correct the mesh for extraneous distortions. When comparing the FEM findings to the experimental data, the friction experienced by each sample lubricant varies at different die strokes, with values ranging from $m = 0.33/\mu = 0.139$ and $m = 0.6/\mu = 0.215$. In comparison to the other benchmark sample, CMFO and NA-O has much higher friction in the first stages I and (ii) ($m = 0.6$; $\mu = 0.215$) than CMFO ($m = 0.5/\mu = 0.1875$) respectively. Nevertheless, when $m = 0.42$ *or* $\mu = 0.1675$, both sample tests show a substantially similar pattern to that of CFC and TSF throughout the remaining phases up to the final product. At stage I, when the CFC and TSF are $m = 0.5/\mu = 0.1875$, the friction for PKO and PMO follows the same trend, and from stage (ii) forward, the load is probably sheared at $m = 0.39/\mu = 0.159$ for both sample lubricants, until completely deformed. In contrast, PS has the smallest measured load value, and its best-fitting value, according to the finite element method, is at $m = 0.5/\mu = 0.1875$, with the trend for the remainder of the stage being best fitted at $m = 0.33/\mu = 0.139$.

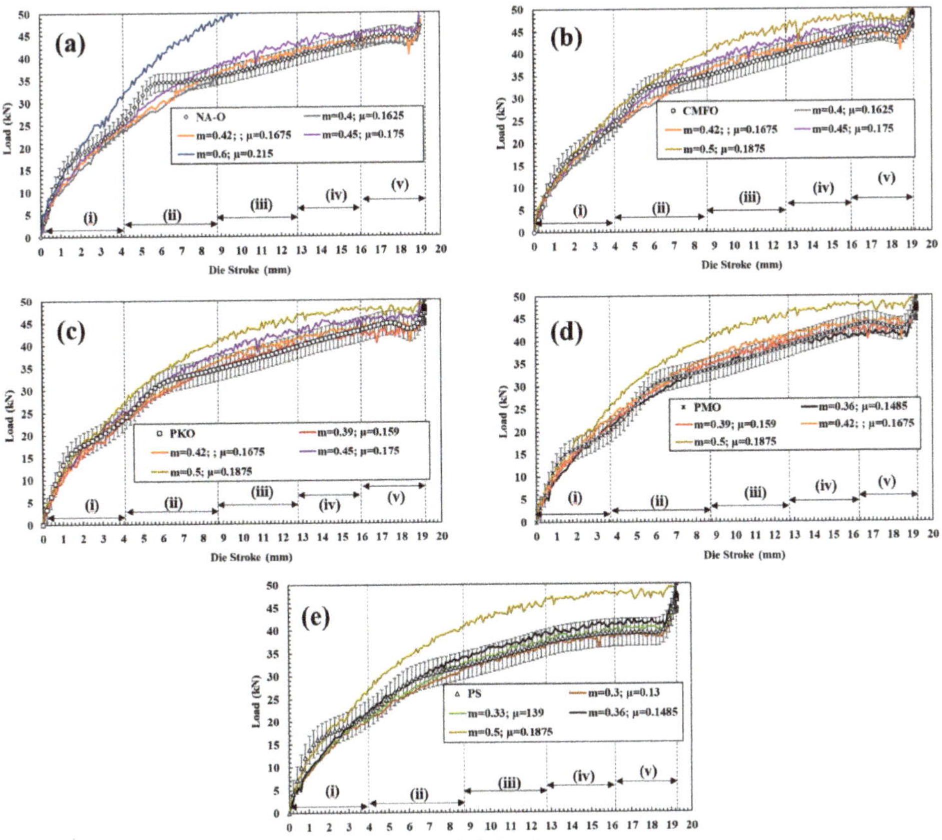

Figure 9. Verification of shear friction coefficient at varying die stroke levels for (**a**) NA-O, (**b**) CMFO, (**c**) PKO, (**d**) PMO and (**e**) PS [15].

After going through the process of optimization, we have come to the conclusion that the development of friction during a closed forging test may be summed up as shown in Figure 10. It appears that the friction behaves differently depending on the die stroke that is being compressed, which is in contrast to the ring compression test, in which the friction may be easily described as a single friction behavior. According to the data, PS had the least amount of friction during the cold forging test when compared to PKO, PMO, and CMFO. PS exhibited a decreasing trend from 10% to 40% as the die stroke compression increased, and this tendency persisted until the compression was complete. Unlike PS, PMO and PKO have a slightly different tendency, in which the friction increases when the compression reaches between 40% and 50%. Another prospective palm oil-based lubricant that may be used as a metal-forming lubricant is called PMO where in comparison to CMFO, the total friction that occurs during the compression test is somewhat lower. PKO, on the other hand, demonstrates practically the same performance as CMFO in terms of friction, which tops the list when compared to other lubricants made from palm oil.

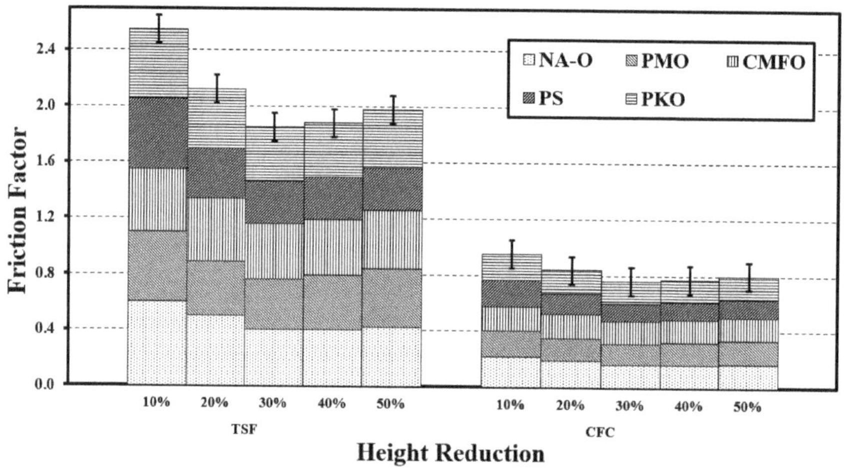

Figure 10. Summary of friction behavior for each component of the sample test following FEM optimization.

It is possible that the fatty acid chain in palm oil, which is used as a metal-forming lubricant, is responsible for its excellent performance. This chain is able to reduce the contact impact between the workpiece and the packed alkyl chain, which reacts with the cumulative short range of van der Waals forces that exist between neighboring groups [27]. A greater amount of closed-packed material led to improved affinity on the metal surface where the unsaturated fatty acid has a double bond on its ninth and tenth carbon chain, which separates it from saturated fatty acids. It was proven by Campen et al. [28] that this double bond was responsible for the formation of cis-configuration in the unsaturated oleic acid, which bent the molecules and made it difficult for them to adopt a linear molecular structure. Because of this, the unsaturated fatty acid is advantageously less efficient in the formation of close-packed monolayer soap film where a lower closed-packed density causes fatty acid chain molecules to have a lower affinity for the surfaces of metals. Meanwhile, the molecules of palmitic and stearic acid, which are saturated fatty acids, have a great capacity to pack tightly and effectively on metal surfaces [34,35].

PS PMO and PKO with high contents of saturated fatty acids displayed effective molecular packing, which protected the metal surface and suggested a reduced friction coefficient [36]. According to Zulhanafi et al. [36], a higher degree of unsaturated fatty acid lowers the capacity of the molecules to be chemisorbed; as a consequence, the metal surfaces in the fourball test experience less protection. However, during the metal-forming process, the lowest level of unsaturated fatty acid, which is PKO (17.7%), has less effect on lowering friction, when compared to PS (46.40%) and PMO (39.29%). This may be the result of very high compression on the surfaces, which reduces the strength of the intermolecular bonds and destroys the thin layer of soap film, resulting in higher friction.

4. Conclusions

In this study, commercial finite element software (DEFORM-3D) was used to simulate two forging processes of aluminum (AA1100) using different compounds of palm oil as a bio-lubricant that has been matched to CMFO. The goals of this study were to analyze the behavior of the flow of material and to predict the forging load and the stress–strain distribution during the forging process.

The results of the open and closed forging tests were shown to have significantly different patterns of frictional behavior. The interpretation of friction in open forging tests is acting as a single friction for each sample test. On the other hand, the behavior of friction in closed forging tests is different at each die stroke, and the friction varies depending on how the stroke is changed.

Both of the findings from the tribological performance of palm oil-based lubricants as metal-forming lubricants demonstrate that palm oil-based lubricants exhibit a superior performance when compared to CMFO. PS has the lowest friction for both tests where the friction in open forging is estimated at $m = 0.10/\mu = 0.05$ and in closed forging process the average is estimated at $m = 0.352/\mu = 0.17$. In open forging test also, PS shows the highest efficiency, which decreased the friction by around 77.8%, while the improvement for the PMO was approximately 66.7%. PKO, on the other hand, demonstrates a worse performance in comparison to PS and PMO, which is somewhere about 55.6%.

In light of the results of this research, one intriguing method for subsequent experiments is the modification of the lubricant, such as employing nano-lubricants in palm oil. This is because the lubrication performance requires the inclusion of additives in order to reduce the friction and improve surface protection.

Author Contributions: The idea was created by A.Y. and S.S., A.Y. was the one who came up with the hypothesis and carried out the calculations. S.S., U.A. and M.K.A.H. checked the validity of the analytical procedures. The manuscript was written by A.Y., who had assistance from U.A. and M.K.A.H. Every contributor contributed insightful comments and suggestions, which were used to improve the study, the analysis, and the manuscript. All authors have read and agreed to the published version of the manuscript.

Funding: This research received no external funding.

Data Availability Statement: There is no data availability.

Acknowledgments: The authors would like to express their thanks to the Ministry of Higher Education of Malaysia for the Konsortium Kecemerlangan Penyelidikan 2021 Research Grant (4L961) and Universiti Teknologi Malaysia (UTM) for the UTM Fundamental Research Grant (22H46).

Conflicts of Interest: The authors declare no conflict of interest.

References

1. Hassan, M.; Ani, F.N.; Syahrullail, S. Tribological performance of refined, bleached and deodorised palm olein blends bio-lubricants. *J. Oil Palm Res.* **2016**, *28*, 510–519. [CrossRef]
2. Yahaya, A.; Samion, S.; Ahyan, N.A.M.; Hamid, M.K.A. Cold extrusion using biodegradable oil as lubricant: Experimental and simulation analysis. *J. Tribol.* **2021**, *30*, 116–132.
3. Quinchia, L.A.; Delgado, M.A.; Franco, J.M.; Spikes, H.A.; Gallegos, C. Low-temperature flow behaviour of vegetable oil-based lubricants. *Ind. Crops Prod.* **2012**, *37*, 383–388. [CrossRef]
4. Syahrullail, S.; Kamitani, S.; Nakanishi, K. Experimental evaluation of refined, bleached, and deodorized palm olein and palm stearin in cold extrusion of aluminum A1050. *Tribol. Trans.* **2012**, *55*, 199–209. [CrossRef]
5. Golshokouh, I.; Syahrullail, S.; Ani, F.N.; Masjuki, H.H. Investigation of palm fatty acid distillate as an alternative lubricant of petrochemical based lubricants, tested at various speeds. *Int. Rev. Mech. Eng.* **2013**, *7*, 72–80.
6. Zulhanafi, P.; Syahrullail, S.; Ahmad, M.A. The tribological performance of hydrodynamic journal bearing using bio-based lubricant. *Tribol. Ind.* **2020**, *42*, 278. [CrossRef]
7. Kramer, P.; Groche, P. Friction measurement under consideration of contact conditions and type of lubricant in bulk metal forming. *Lubricants* **2019**, *7*, 12. [CrossRef]

8. Okokpujie, I.P.; Tartibu, L.K.; Sinebe, J.E.; Adeoye, A.O.; Akinlabi, E.T. Comparative Study of Rheological Effects of Vegetable Oil-Lubricant, TiO_2, MWCNTs Nano-Lubricants, and Machining Parameters' Influence on Cutting Force for Sustainable Metal Cutting Process. *Lubricants* **2022**, *10*, 54. [CrossRef]
9. Moshkovich, A.; Perfilyev, V.; Rapoport, L. Effect of plastic deformation and damage development during friction of FCC metals in the conditions of boundary lubrication. *Lubricants* **2019**, *7*, 45. [CrossRef]
10. Hafis, S.M.; Ridzuan, M.J.M.; Farahana, R.N.; Ayob, A.; Syahrullail, S. Paraffinic mineral oil lubrication for cold forward extrusion: Effect of lubricant quantity and friction. *Tribol. Int.* **2013**, *60*, 111–115. [CrossRef]
11. Groche, P.; Kramer, P.; Bay, N.; Christiansen, P.; Dubar, L.; Hayakawa, K.; Hu, C.; Kitamura, K.; Moreau, P. Friction coefficients in cold forging: A global perspective. *CIRP Ann.* **2018**, *67*, 261–264. [CrossRef]
12. Zhang, D.W.; Ou, H. Relationship between friction parameters in a Coulomb–Tresca friction model for bulk metal forming. *Tribol. Int.* **2016**, *95*, 13–18. [CrossRef]
13. Du, F.; Li, C.; Li, D.; Sa, X.; Yu, Y.; Li, C.; Yang, Y.; Wang, J. Research Progress Regarding the Use of Metal and Metal Oxide Nanoparticles as Lubricant Additives. *Lubricants* **2022**, *10*, 196. [CrossRef]
14. Sani, A.; Sahab, A.; Abd Rahim, E.; Talib, N.; Kamdani, K.; Rahim, M.Z. Performance Evaluation of Palm-Olein TMP Ester Containing Hexagonal Boron Nitride and an Oil Miscible Ionic Liquid as Bio-Based Metalworking Fluids. *J. Mech. Eng.* **2017**, *4*, 223–234.
15. Aiman, Y.; Syahrullail, S.; Hamid, M.K.A. Optimisation of friction surfacing process parameters for a1100 aluminium utilising different derivatives of palm oil based on closed forging test. *Biomass Convers. Biorefin.* **2022**, 1–18. [CrossRef]
16. Stoffel, W.; Chu, F.; Ahrens, E.H. Analysis of long-chain fatty acids by gas-liquid chromatography. *Anal. Chem.* **1959**, *31*, 307–308. [CrossRef]
17. Aiman, Y.; Syahrullail, S. Frictional and material deformation of aluminium alloy in cold forging test under different derivatives of palm oil lubrication condition. *J. Braz. Soc. Mech. Sci. Eng.* **2022**, *44*, 396. [CrossRef]
18. Abdulmawlla, M.A. Finite Element Analysis and Optimization of Closed Die Forging Process for Aluminium Metal Matrix Composites. Ph.D. Thesis, Universiti Putra Malaysia, Serdang, Malaysia, 2010.
19. Okokpujie, I.P.; Chima, P.C.; Tartibu, L.K. Experimental and 3D-Deform Finite Element Analysis on Tool Wear during Turning of Al-Si-Mg Alloy. *Lubricants* **2022**, *10*, 341. [CrossRef]
20. Li, F.; Chen, P.; Han, J.; Deng, L.; Yi, J.; Liu, Y.; Eckert, J. Metal flow behavior of P/M connecting rod preform in flashless forging based on isothermal compression and numerical simulation. *J. Mater. Res. Technol.* **2020**, *9*, 1200–1209. [CrossRef]
21. Patwardhan, P.S.; Nalavde, R.A.; Kujawski, D. Estimation of Ramberg-Osgood Constants for Materials with and without Luder's Strain Using Yield and Ultimate Strengths. *Procedia Struct. Integr.* **2019**, *17*, 750–757. [CrossRef]
22. Szala, M.; Winiarski, G.; Wójcik, Ł.; Bulzak, T. Effect of Annealing Time and Temperature Parameters on the Microstructure, Hardness, and Strain-Hardening Coefficients of 42CrMo4 Steel. *Materials* **2020**, *13*, 2022. [CrossRef]
23. Tiong, C.I.; Azli, Y.; Kadir, M.R.A.; Syahrullail, S. Tribological evaluation of refined, bleached and deodorized palm stearin using four-ball tribotester with different normal loads. *J. Zhejiang Univ. Sci. A* **2012**, *13*, 633–640. [CrossRef]
24. Wang, J.P. A new evaluation to friction analysis for the ring test. *Int. J. Mach. Tools Manuf.* **2001**, *41*, 311–324. [CrossRef]
25. Farhanah, A.N.; Syahrullail, S.; Ajruddin, M.A. Evaluation of RBD palm stearin as alternative lubricant for cold forward extrusion process. *J. Phys. Conf. Ser.* **2017**, *908*, 012062. [CrossRef]
26. Maleque, M.A.; Masjuki, H.H.; Sapuan, S.M. Vegetable-based biodegradable lubricating oil additives. *Ind. Lubr. Tribol.* **2003**, *55*, 137–143. [CrossRef]
27. Crespo, A.; Morgado, N.; Mazuyer, D.; Cayer-Barrioz, J. Effect of Unsaturation on the Adsorption and the Mechanical Behavior of Fatty Acid Layers. *Langmuir* **2018**, *34*, 4560–4567. [CrossRef]
28. Campen, S.; Green, J.H.; Lamb, G.D.; Spikes, H.A. In situ study of model organic friction modifiers using liquid cell AFM; saturated and mono-unsaturated carboxylic acids. *Tribol. Lett.* **2015**, *57*, 18. [CrossRef]
29. Zhang, D.W.; Cui, M.C.; Cao, M.; Ben, N.Y.; Zhao, S.D. Determination of friction conditions in cold-rolling process of shaft part by using incremental ring compression test. *Int. J. Adv. Manuf. Technol.* **2017**, *91*, 3823–3831. [CrossRef]
30. Caminaga, C.; da Silva Issii, R.L.; Button, S.T. Alternative lubrication and lubricants for the cold extrusion of steel parts. *J. Mater. Process. Technol.* **2006**, *179*, 87–91. [CrossRef]
31. Nurul, M.A.; Syahrullail, S. Lubricant viscosity: Evaluation between existing and alternative lubricant in metal forming process. *Procedia Manuf.* **2015**, *2*, 470–475. [CrossRef]
32. Yingying, L.; Houfang, L.; We, J.; Dongsheng, L.; Shijie, L.; Bin, L. Biodiesel Production from crude *Jatropa curcas* L. Oil with Trace Acid Catalyst. *Chin. J. Chem. Eng.* **2012**, *20*, 740–746.
33. Liu, Y.; Lu, H.; Jiang, W.; Li, S.; Liu, S.; Liang, B. Biodiesel production from crude *Jatropha curcas* L. oil with trace acid catalyst. *Chin. J. Chem. Eng.* **2012**, *20*, 740–746. [CrossRef]
34. Razak, D.M.; Syahrullail, S.; Sapawe, N.; Azli, Y.; Nuraliza, N. A new approach using palm olein, palm kernel oil, and palm fatty acid distillate as alternative biolubricants: Improving tribology in metal-on-metal contact. *Tribol. Trans.* **2015**, *58*, 511–517. [CrossRef]

35. Wood, M.H.; Casford, M.T.; Steitz, R.; Zarbakhsh, A.; Welbourn, R.J.L.; Clarke, S.M. Comparative adsorption of saturated and unsaturated fatty acids at the iron oxide/oil interface. *Langmuir* **2016**, *32*, 534–540. [CrossRef]
36. Zulhanafi, P.; Syahrullail, S. The tribological performances of Super Olein as fluid lubricant using four-ball tribotester. *Tribol. Int.* **2019**, *130*, 85–93. [CrossRef]

lubricants

MDPI

Article

Boundary Lubricity of Vegetable-Oil-Derived Trimethylolpropane (TMP) Ester

Chiew Tin Lee [1], Mei Bao Lee [1], William Woei Fong Chong [1,2,*], Jo-Han Ng [3], King Jye Wong [1] and Cheng Tung Chong [4]

[1] Faculty of Mechanical Engineering, Universiti Teknologi Malaysia (UTM), Johor Bahru 81310, Johor, Malaysia; ctlee5@graduate.utm.my (C.T.L.); mblee2@graduate.utm.my (M.B.L.); kjwong@utm.my (K.J.W.)
[2] Automotive Development Centre (ADC), Institute for Vehicle Systems and Engineering (IVeSE), Universiti Teknologi Malaysia (UTM), Johor Bahru 81310, Johor, Malaysia
[3] Faculty of Engineering and Physical Sciences, University of Southampton Malaysia, Iskandar Puteri 79100, Johor, Malaysia; j.ng@soton.ac.uk
[4] China-UK Low Carbon College, Shanghai Jiao Tong University, Shanghai 201306, China; ctchong@sjtu.edu.cn
* Correspondence: william@utm.my

Abstract: Vegetable-oil-based biolubricants are an excellent alternative to conventional lubricants. Instead of focusing on novel feedstocks, these biolubricants should be further elucidated based on their fatty acid composition, which influences their tribological properties. Therefore, the study utilises gene expression programming (GEP) to derive a boundary lubricity model for vegetable-oil-derived trimethylolpropane (TMP) esters, considering the fatty acid composition (saturation and monounsaturation levels), load and speed. Neat vegetable oil and blends from seven feedstocks are selected following a wide range of fatty acid profiles to synthesise TMP esters using a two-stage transesterification process. The TMP esters are spin-coated on wear discs that are subsequently rotated against a ball using a purpose-built tribometer. The frictional performance of the TMP esters with balanced saturation and monounsaturation levels of fatty acid are measured to improve it at higher speeds. The GEP model is statistically evaluated by adopting the friction data, a showing good generalisation and predictability capability. The model demonstrates that friction decreases with increasing saturation levels of the TMP ester. The GEP model for vegetable oil TMP esters allows for the tribological performance prediction of TMP esters following the fatty acid profile, providing a platform to optimise such biolubricant for desired applications.

Keywords: fatty acid composition; tribometer; spin coating; thin film; physicochemical; gene expression programming

1. Introduction

The global biolubricant market was estimated at USD 1.9 billion in 2020 and is projected to reach USD 2.7 billion by the year 2027 [1]. The estimated growth in the biolubricant sector at a compound annual growth rate (CAGR) of 5.2% is driven by concerns about the rapid depletion of fossil fuel resources and environmental pollution arising from improper disposal of lubricants derived from such resources. Biolubricants are an alternative to conventional lubricants derived from fossil fuel resources. They are biodegradable (70–100% [2]) and renewable, with the most common type being vegetable oil or its derivatives (approximately 45% of the global biolubricant market [1]). In addition to not being toxic, vegetable oil also has excellent lubrication properties, attributed to its unique combination structure of polar and nonpolar molecular groups. The carboxyl polar group of vegetable oil adsorbs to the rubbing surfaces to form a lubricant film, protecting them from undesired wear and tear.

Vegetable oil's high fatty acid content (between 80–90%) is the primary factor in its lubricating efficacy [3]. However, the β-hydrogen atoms in the hydroxyl groups and unsaturated free fatty acids (FFA) of the vegetable oil triglycerides promote fast crystallisation, resulting in poor thermo-oxidative stability. Therefore, converting vegetable oil into synthetic esters through chemical modifications (e.g., epoxidation, hydrogenation and transesterification [4]) can remedy this deficiency. One of the economically feasible solutions is to convert vegetable oil into a synthetic polyol ester, namely, trimethylolpropane ester (TMP ester), using transesterification. On top of an improved thermo-oxidative stability [4,5], numerous research studies on such polyol esters derived from vegetable oil have produced biolubricants with a lubricity that is on par or superior to conventional lubricants [6,7]. To date, vegetable oils, such as palm oil [8], jatropha oil [9], rice bran [4], karanja oil [4], sunflower oil [9], soybean oil [9,10] and cotton seed oil [11], have been transesterified to produce TMP esters. Recently, studies on TMP esters have shifted towards specific fatty acid chains, namely, from TMP trioleate [12,13]. TMP trioleate, synthesised from oleic acid, has been demonstrated to possess a good lubricity and has often been suggested for use as hydraulic oil.

A recent bibliometric study by Lee et al. [14] highlighted the importance of correlational studies on the lubricity of vegetable oil biolubricant and its fatty acid composition. The chemical structure of vegetable-oil-derived biolubricants and their fatty acid composition is vital to their physicochemical and tribological properties [15–17]. According to the study by Biresaw and Bantchev [18], the lubricant film is influenced by the degree of unsaturation, fatty acid chain length and the polar group of the vegetable oil. Correlating the fatty acid compositions of vegetable-oil-based lubricants with rheological properties has been reported on numerous occasions in the literature [9,19,20]. For example, the monounsaturated or polyunsaturated fatty acids influence the viscosity of vegetable oil [19]. Specifically, Kim et al. demonstrated that the viscosity of vegetable oil increased with a higher concentration of oleic acid (C18:1) while it decreased with a higher linoleic acid (C18:2) content [20]. Apart from viscosity, it has also been reported that a low thermo-oxidative stability is attributed to higher levels of unsaturation [9] while high pour points are attributed to higher levels of saturation [21].

On the other hand, the influence of fatty acid profiles on vegetable oil's frictional and wear properties have also been studied. For example, stearic acid in vegetable-oil-based lubricants could reduce friction and wear [3]. High linoleic and oleic acid concentrations in vegetable oils, namely soybean oil, have also produced a lower friction and wear, arising from the formation of denser fatty acid monolayer film [22,23]. Hamdan et al. suggested that decreasing the ratio of monounsaturation to total saturation level could result in reducing the friction of vegetable-oil-derived fatty acid methyl ester (FAME) [24]. A reducing ratio would also reflect an increasing saturation level (assuming a constant monounsaturation). Similarly, the increasing saturation level of vegetable oil FAME has also been reported to lead to a friction drop by Rajasozhaperumal and Kannan [23]. They explained that saturated fatty acid molecules adsorbed more efficiently on the surface to form a more effective lubricating film.

The literature mentioned above often highlights the effect of the fatty acid composition of vegetable-oil-based lubricants with little effort on mathematically quantifying the effect. Neat vegetable oil feedstocks are often investigated along with different chemical modification and additivation approaches. Instead of using a trial-and-error approach, a proper quantification of such effect would be imperative for optimising the tribological properties of vegetable oil for developing more effective biolubricants, allowing an optimum fatty acid configuration of biolubricant to be attained for the desired applications. Therefore, machine learning techniques should be explored when determining the effect of fatty acid composition on the tribological properties of vegetable-oil-based biolubricants. Machine learning allows complex processes to be systematically quantified in an efficient manner [25], suitable for tribological systems, often involving multiphysics phenomena.

Marian and Tremmel recently reviewed the penetration of machine learning techniques in the field of tribology research [26]. Their review showed that most tribological research adopting machine learning techniques are related to composite or advanced materials, lubrication systems for motion generation or power transmissions (including bearings, seals, brakes and clutches), surface texturing and lubricants. With relation to lubricants, the reported literature focuses more on additive studies [27–29]. Artificial Neural Networks (ANNs) and genetic algorithms are the majority of machine learning techniques adopted in tribology. Machine learning techniques solve the identified problem by training using available data from a simulation, experiment or the literature. To date, ANNs have been demonstrated to produce predictive models with a high coherence to experimental data [30]. However, it remains a challenge to derive an empirical formulation of an ANN model for practical applications due to the complex nature of ANN models [31].

An alternative machine learning technique is gene expression programming (GEP), an extension of genetic programming that incorporates simple and linear chromosomes to generate small programs with explicit equations [32]. GEP has the genetic algorithm's simplicity and the abilities of genetic programming [33]. This technique produces simple mathematical expressions in the form of subexpression trees with high prediction capabilities that can be adopted for practical applications. More importantly, the expressions from GEP have been reported in the literature to have a good generalisation and predictive capability, not limited to correlations [30,34,35]. Recently, GEP has been adopted to study material removal via machining [36], with most studies still revolving around civil engineering applications [30,37].

Knowing the influence of different fatty acid compositions on the physicochemical and tribological properties of vegetable-oil-based biolubricant is essential. A lubricant's fluid film lubrication performance is heavily influenced by its viscosity. However, the same cannot be said of its boundary lubrication properties. Therefore, as a first approximation, the present study aims to develop a practical empirical expression using GEP to describe the effect of vegetable-oil-based trimethylolpropane (TMP) ester's fatty acid composition on its boundary frictional performance. The generated GEP model prepares an empirical platform to further explore the boundary lubricity of TMP esters as an alternative to conventional mineral-oil-based lubricants. More importantly, the GEP model is expected to predict the frictional performance of vegetable-oil-based TMP esters following the operating conditions of the desired applications. To the authors' knowledge, adopting the GEP technique in deriving a generalised empirical model for vegetable-oil-derived TMP esters' boundary frictional performance, considering fatty acid composition, has yet to be reported in the literature.

2. Methodology

2.1. Materials

The present study derived trimethylolpropane (TMP) esters from vegetable oil feedstocks, namely palm, olive, coconut, grapeseed–coconut (1:1 ratio by volume), canola, canola–sunflower (1:1 ratio by volume) and canola–palm–soybean (1:1:1 ratio by volume). It is noted that these vegetable oils were commercially procured from the consumer product market. The methanol and potassium hydroxide (KOH) used in the methyl ester synthesis were of the QRëC brand. TMP: 1,1,1-Tris(hydroxymethyl)propane, dist., ≥98.0% (by gas chromatography) and 30 wt% sodium methoxide solution in methanol were purchased from Sigma-Aldrich (Selangor, Malaysia) for synthesising trimethylolpropane (TMP) ester. All chemicals used in the production were of analytical grade. On the other hand, BSTFA: N,O-Bis(trimethylsilyl)trifluoroacetamide (synthesis grade, purity >99%) and ethyl acetate (ACS reagent grade) were also purchased from Sigma-Aldrich (Malaysia) for the gas chromatography (GC) analysis of the produced TMP ester.

2.2. Vegetable-Oil-Derived Trimethylolpropane (TMP) Ester

The synthesis of TMP esters for the selected vegetable oil feedstocks was carried out through a two-stage transesterification process as described in the process flow diagram given in Figure 1. In the first stage of the transesterification (process flows 1 to 7), the triglyceride of vegetable oil was reacted with methanol in the presence of a base catalyst

(KOH), forming a fatty acid methyl ester (FAME) and glycerol. The collected FAME was then sent for gas chromatography–mass spectrometry (GCMS) using a Shimadzu GCMS-QP2010 Series coupled with a BP5 MS column (20 m × 0.18 mm (inside diameter), film thickness 0.18 μm) to identify their saturation (SA), monounsaturation (MU) and polyunsaturation (PU) levels.

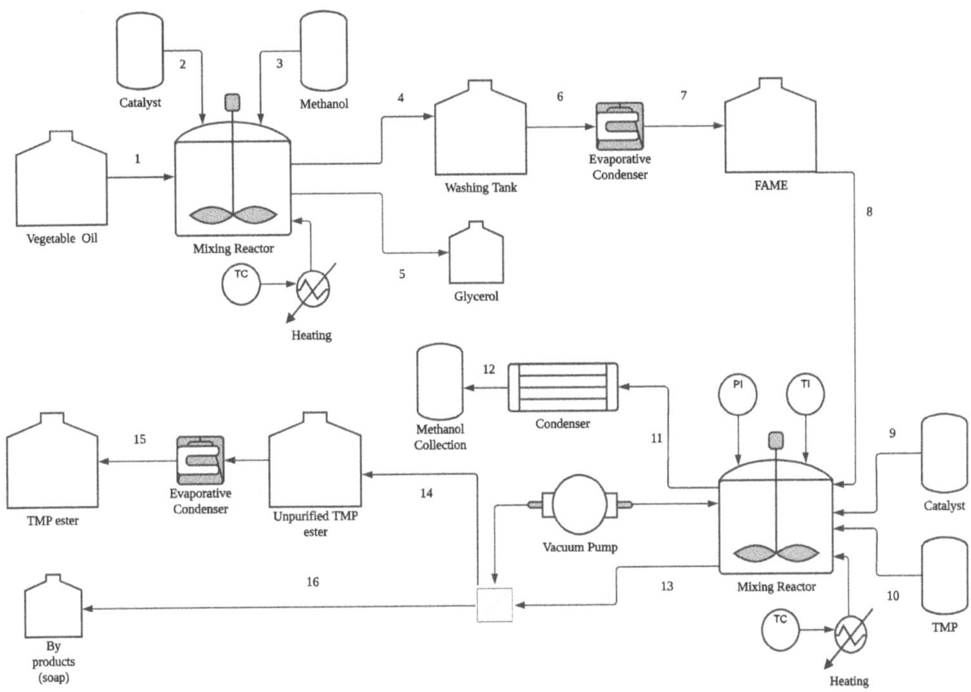

Figure 1. Two-stage transesterification process flow diagram for TMP ester synthesis.

The methodology for GCMS was modified following Tarif et al. [38]. The column temperature was set from 120 °C to 330 °C at rate of 10 °C/min. Both injector and detector temperatures were set at 250 °C. A quantity of 1.0 μL of test samples in ethyl acetate was injected with a split ratio of 1:10. Helium gas was used as the carrier gas at a flow rate of 1.5 mL/min. The mass spectrometers were then identified with the National Institute of Standards and Technology (NIST) mass spectrometer library across a range of 50–550 m/z with an electron impact (EI) ionisation mode. The fatty acid composition for the synthesised FAME is tabulated in Table 1. The present study accepted only a distribution yield of at least 96.5% (following the EN14214 standard) from the FAME for each vegetable oil to ensure a high conversion of the TMP ester production in the subsequent transesterification process.

In the second stage of transesterification (process flows 8 to 15 in Figure 1), the FAME was reacted with TMP in the presence of sodium methoxide as the base catalyst in reflux and vacuum conditions at 130 °C. After five hours of reaction time, the reactant fluid was vacuum filtered using a Whatman Grade 5 qualitative filter paper to remove all unused by-products. Finally, the filtrated reactant fluid underwent distillation to remove the excess FAME, leaving behind the desired TMP ester. The two-stage transesterification allowed the β-hydrogen in the glycerol of the vegetable oil to be replaced by TMP, conjugating the FAME into a synthetic ester.

The quantification of the yield of TMP esters was then conducted through a gas chromatography (GC) system (Perkin Elmer Clarus 680) equipped with a flame ionisation detector (FID). The GC instrument was operated on an Elite-5 capillary column (30 m × 0.25 mm (inside diameter), with a film thickness of 0.25 μm). The GC technique followed the methodology from Yunus et al. [39]) with minor modifications to suit the column used. A (0.03 ± 0.005) g of the TMP ester was weighed and diluted with 1.0 mL of ethyl acetate and 0.5 mL of BSTFA. The samples were heated in a water bath at 40 °C for silylation. During the GC analysis, hydrogen gas was employed as a carrier gas with a split ratio of 50:1. The oven temperature was first held at 80 °C for three minutes initially and heated up to 330 °C at 5 °C/min, before being held for another eight minutes. The injector and detector temperatures were set at 300 °C and 330 °C, respectively.

Table 1. Fatty acid composition of vegetable oils.

Parameter	Palm	Olive	Coconut	Grapeseed–Coconut	Canola	Canola–Sunflower	Canola–Palm–Soybean
C8:0	0.02	0.07	3.89	3.79	-	0.13	0.09
C10:0	0.01	0.10	5.16	4.82	0.01	0.19	0.09
C12:0	0.13	1.03	19.30	14.99	0.03	1.81	0.98
C14:0	1.51	0.55	12.89	9.93	0.02	1.08	2.05
C16:0	34.47	16.83	10.16	10.40	0.04	11.05	25.22
C18:0	6.51	4.99	4.37	2.58	0.12	5.73	6.79
C20:0	-	1.92	0.36	1.43	0.09	1.70	1.57
C21:0	-	0.36	-	0.40	2.28	-	0.06
C16:1	0.51	4.53	0.09	1.31	0.69	0.83	0.75
C18:1	48.05	47.83	13.21	2.07	11.31	15.21	13.42
C20:1	0.81	1.49	0.70	3.01	2.27	2.98	1.08
C18:2	0.38	-	-	-	73.83	2.56	1.42
C18:3	0.33	4.27	3.56	26.75	5.21	44.93	39.21
Other	7.27	16.04	26.30	18.52	4.10	11.78	7.27
Saturation, SA (%)	46.00	30.79	76.17	59.32	2.70	24.60	39.74
Monounsaturation, MU (%)	53.24	64.13	19.00	7.85	14.88	21.57	16.44
Polyunsaturation, PU (%)	0.76	5.08	4.83	32.83	82.42	53.83	43.82

The number of peaks in the gas chromatogram (GC) is related to the concentration of a specific component in the test samples. The GC peaks illustrate the number of carbon atoms bound in an ester structure. The present study used methyl oleate (MO) and trimethylolpropane oleate (TMPO) to identify the corresponding C18 bond in both MO and TMPO. MO with a ≥98.5% purity was purchased from Sigma-Aldrich (Selangor, Malaysia), while Wilmar Oleochemicals (Jiangsu, China) provided the TMPO. The respective chromatograms are given in Figure 2a,b. The ester formation for the produced TMP ester can be recognised by comparing the GC peaks with MO and TMPO. For example, Figure 2c illustrates the chromatogram for the palm TMP ester. MO's most prominent GC peak occurs at 27.5 min, identified as methyl oleate. On the other hand, the highest GC peak in TMPO is at 57.7 min, identified as trimethylolpropane oleate. In comparison, the GC peaks of the palm TMP ester appear around 57 min. These peaks demonstrate the combination of TMP esters formed from the C18 methyl ester families (C18:0. C18:1, C18:2 and C18:3). In the present study, the synthesis of TMP esters was optimised to ensure a TMP ester yield distribution of at least 85% for each vegetable oil. For example, by the weight percentage calculation, the palm TMP ester produced was estimated to contain 87% of TMP ester. It is noted that the produced TMP esters were also subjected to distillation to remove any residue or remaining FAME before further characterisation.

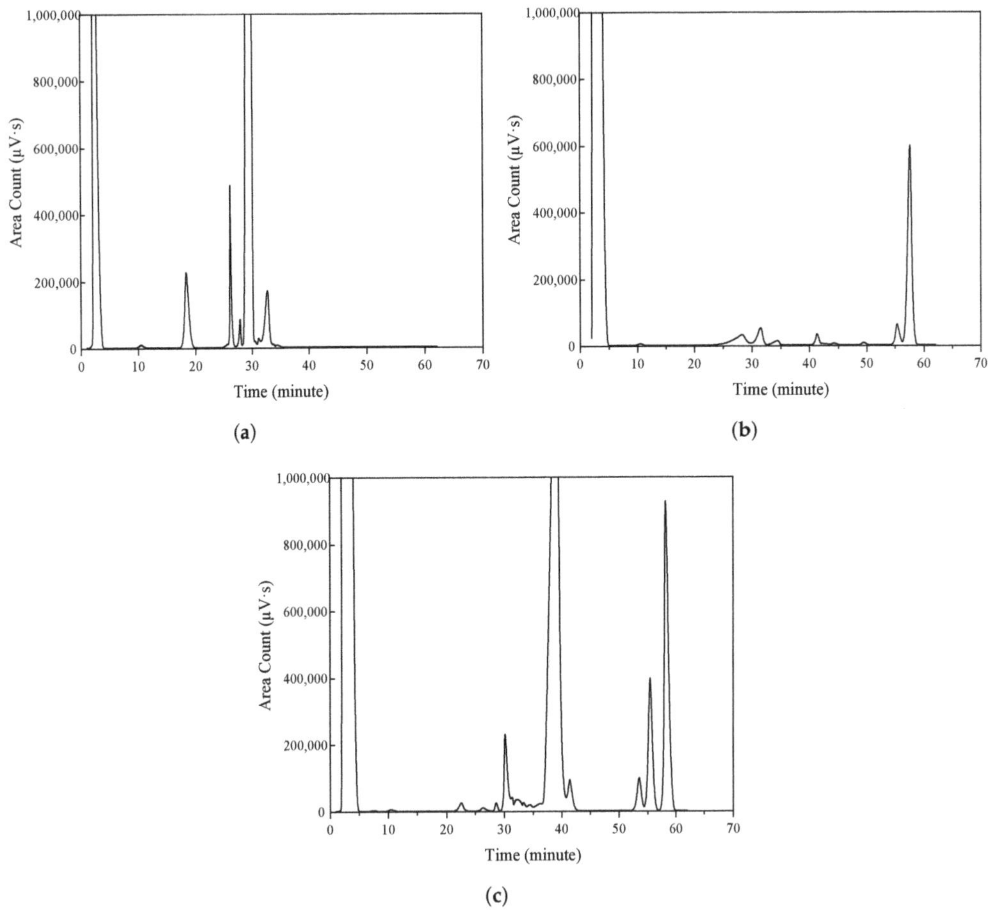

Figure 2. Example gas chromatogram for palm TMP ester. (**a**) Methyl oleate (MO), (**b**) trimethylolpropane oleate (TMPO), (**c**) palm TMP ester.

2.3. *Physicochemical Properties of Trimethylolpropane (TMP) Esters*

The viscosities of the derived TMP esters were measured using a Brookfield DV-11+ Pro viscometer, equipped with a constant-temperature bath to control the temperature of the tested liquid at 40 °C and 100 °C. The kinematic viscosity and viscosity index (VI) were determined following ASTM D445 and ASTM D2270, respectively. The flash points of the TMP esters were also measured based on ASTM D93-94, using a Pensky–Martens closed-cup tester (Anton Paar, Graz, Austria). At the same time, the pour point was measured according to ASTM D97-93 (Stanhope-seta). The density was determined using the Anton Paar DMA 4100. Lastly, the thermogravimetric analysis (TGA) was carried out using a TA Instruments (New Castle, DE, USA) Thermogravimetric Analyzer (modular TGA Q500) in a temperature range from room temperature to 1000 °C in a nitrogen gas atmosphere at a 10 °C/min heating rate to determine the thermal decomposition onset temperature of the TMP esters.

2.4. *Spin Coating of Lubricant Film*

In the present study, a dynamic spin-coating method was adopted to deposit a thin layer of TMP ester on a 304 2B finished stainless steel wear disc (50 mm diameter and

1 mm thickness). The stainless steel wear discs were first rinsed with water, followed by sonication in acetone to remove residual machining or cutting fluid, before being washed with distilled water. The discs were dried in the oven and kept in ethanol as a pretreatment to improve their adhesion properties. Subsequently, the stainless steel wear discs were removed from the ethanol and dried in the oven at 80 °C for 30 min to evaporate the ethanol before being cooled to room temperature. During spin coating, 0.5 mL of TMP ester was added dropwise to the middle of the stainless steel wear discs. The centripetal acceleration was set at 3000 rev/min, spreading the TMP ester while forming a thin layer of lubricant film on the surface of the wear discs. The wear discs coated with TMP ester were weighed with the thickness of the coated film being evaluated using the mass difference relative to the uncoated discs, following reference [40]. It is noted that the spin-coating approach was adopted to allow for the characterisation of the TMP ester at a boundary lubrication regime.

2.5. Friction Measurement

The friction test was conducted under ambient conditions with an in-house purpose-built ball-on-disc tribometer, given in Figure 3. The purpose-built ball-on-disc tribometer used a high-speed brushless DC motor, allowing a rotational speed of up to 6000 rev/min for the wear disc with a gear reducer. The friction was measured using a calibrated load cell, while the rotational speed was measured using a Hall effect sensor. The 304 2B finished stainless steel wear disc (spin-coated with TMP esters) was rotated against a steel ball of 6 mm diameter (fixed at 1.75 cm from the centre of the discs) at varied normal loads, w (392–1177 mN, giving an estimated maximum Hertzian pressure of 0.5–0.7 GPa) and rotating speeds, u (0.36–7.33 m/s). The friction for each test configuration was measured for 90 s. Before the friction test, the steel ball was cleaned with acetone while a new set of steel balls and wear discs was used for each TMP ester. The wear discs' arithmetic average roughness (Ra) and root mean square average roughness (Rq) were 1.25 μm and 1.61 μm, respectively. The Ra and Rq for the ball bearing were 0.2 μm and 0.3 μm, respectively.

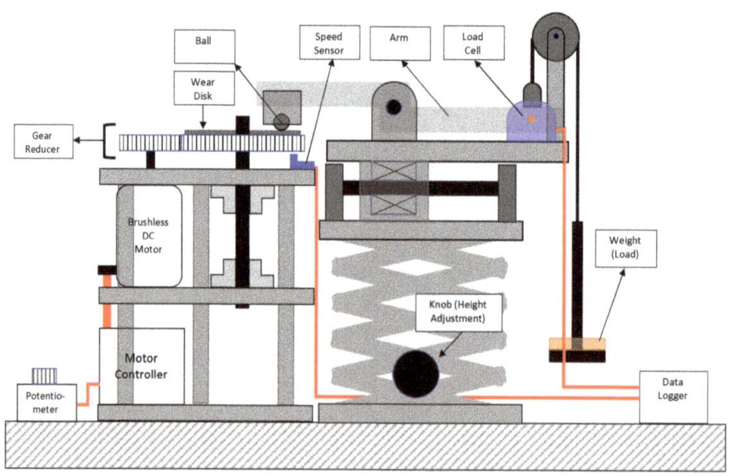

Figure 3. Purpose-built ball-on-disc tribometer.

The lambda parameter (λ) was estimated from the ratio between the minimum film thickness (h_{min}) and the composite surface roughness (σ), allowing for the interpretation of the operating lubrication regime. The minimum film thickness (h_{min}) was determined at each operating condition following the expression by Hamrock and Dowson [41], while the term σ was the composite surface roughness of the wear disc and ball bearing. The expression for h_{min} was given as:

$$\frac{h_{min}}{R} = 3.63U^{0.68}G^{0.49}W^{-0.073}\left(1 - e^{-0.68}\right) \quad (1)$$

where:
$G + \alpha E'$;
$U = u\eta_0 / E'R$;
$W = w / E'R^2$;
R = curvature radius of ball;
E' = reduced modulus of elasticity; $\left(= 2\left[(1 - \nu_1^2)/E_1 + (1 - \nu_2^2)/E_2\right]^{-1}\right)$
E = modulus of elasticity for wear disc and ball;
ν = Poisson's ratio for wear disc and ball;
w = load;
u = sliding speed;
η_0 = dynamic viscosity.

2.6. Gene Expression Programming (GEP)

The present study adopted a gene expression programming (GEP) approach to determine a meaningful relationship between the friction force (F_f) and the fatty acid composition of the vegetable-oil-derived TMP esters. The variables identified were saturation (SA), monounsaturation (MU), polyunsaturation (PU), load and speed. The interdependency of variables could result in models with poor performance, where the strength between these variable pairs could be exaggerated due to multicollinearity [42,43]. It is noted that the average carbon chain lengths were excluded as an input variable due to their marginal difference, where the lengths of the selected vegetable-oil-derived TMP esters were mostly 18, except for coconut (15) and grapeseed–coconut (16) TMP esters. Therefore, in the present study, the multicollinearity of variables was identified using the correlation coefficient between the variables as in Table 2. The PU–SA and PU–MU variable pairs gave relatively higher correlation coefficients than other variable pairs. Thus, as a first approximation, the variable PU was removed from the GEP model generation, giving a function as follows:

$$F_f = f(SA, MU, load, speed) \quad (2)$$

An acceptable machine-learning-based model is suggested to have a ratio of total data points to total input variables of at least three, preferably exceeding five [34]. In the present study, the GEP database was formed from 175 values measured through friction tests, giving a ratio of more than 40 with four input variables. Then, the database was randomly distributed into training, validation and test subsets using a stratified sampling technique. A total of 141 (80%) values were taken for the GEP training process. The remaining values were equally divided for the validation (17 values—10%) and test (17 values—10%) subsets required to validate the GEP model's generalisation capability. For every 2000 generations during GEP, the complexity was increased by automatically adding a gene to the model. A parametric study was conducted on the number of chromosomes, head size and the number of genes using the training data to achieve a proper GEP architecture. The optimum configuration based on best fitness and correlation coefficient for the current data set could be obtained using 225 chromosomes, a head size of 10 and 6 genes. Appendix A provides the input parameters for the GEP algorithm using GeneXproTools Software (Version 5.0). It is noted that the GEP in the present study was terminated at 20,000 generations.

Table 2. Correlation coefficient matrix for variables

Parameters	SA (%)	MU (%)	PU (%)	Load (mN)	Speed (m/s)
SA (%)	1				
MU (%)	−0.103	1			
PU (%)	−0.715	−0.621	1		
Load (mN)	−0.027	−0.047	0.054	1	
Speed (m/s)	0.058	0.033	−0.069	−0.043	1

The performance evaluation of the GEP model was based on the statistical parameters for training, validation and test subsets. The parameters considered were the root-mean-square error ($RMSE$), mean absolute error (MAE), relative standard error (RSE), relative root-mean-square error ($RRMSE$), correlation coefficient (R) and determination coefficient (R^2). A performance index, ρ, was introduced following Gandomi and Roke [34]. On top of these parameters, the present study also conducted an external validation to evaluate the prediction capability of the GEP model. The parameters considered were the slope of the regression lines (k and k'), the squared correlation coefficient through the origins ($R_o'^2$) and the coefficient between the experiment and model (R_o^2) [30]. Overfitting models due to excessive data training could result in a decreased training error but an increased testing error [44]. Therefore, an objective function (OBF) parameter was used to measure the overall performance of the GEP model. The equations for these statistical parameters are given in Appendix B.

3. Results and Discussion

Table 3 summarises the physicochemical properties of the vegetable-oil-derived TMP esters. Based on the kinematic viscosity values, the palm and grapeseed–canola TMP esters fell under the ISO VG 22 grade. On the other hand, the olive and canola–palm-soybean TMP esters could be classified under ISO VG 68, while canola and canola–sunflower followed ISO VG 150. Contrary to the observation by Kim et al. [20], instead of a lower viscosity, higher viscosity values for canola and canola–sunflower could result from the coupled effect of the high mono- and polyunsaturation levels as shown in Table 1. On the other hand, the derived coconut TMP ester produced the lowest viscosity, attributed to its shorter alkyl carboxylic chain (predominantly C12 to C16). At the present state, this TMP ester also could not be satisfactorily classified as its viscosity fell between ISO VG 10 and 15. However, the coconut TMP ester could be modified with suitable viscosity modifiers to suit either viscosity grade.

The viscosity index (VI) values for most of the derived TMP esters were calculated to be higher or comparable to reported typical values of oil-based engine lubricants. The canola and canola–sunflower TMP esters' VI values were lower among the derived TMP esters at 121 and 132, respectively. By referring to Table 1, the lower VI values from these TMP esters could be attributed to their high PU levels (>50%). The highest VI value was recorded by the coconut TMP ester at 259, potentially resulting from its high SA levels. Such a high VI value shows potential for the coconut TMP ester to be adopted for systems operating under harsh conditions. The thermal decomposition onset temperature determined using TGA can indicate the lubricant's thermal stability. In the present study, all the derived TMP esters exhibited a decomposition onset temperature above 300 °C, which was much higher than the values reported for engine lubricants (230 to 260 °C) [45].

Table 3. Physicochemical properties and spin-coated film thickness of vegetable-oil-derived TMP esters.

Parameter	**Palm**	**Olive**	**Coconut**	**Grapeseed–Coconut**	**Canola**	**Canola–Sunflower**	**Canola–Palm–Soybean**
Density (g/mL) @25 °C	0.902	0.921	0.898	0.954	0.935	0.937	0.922
Kin. viscosity (mm/s^2) @40 °C	22.95	72.29	12.87	20.82	154.50	142.16	65.34
@100 °C	5.45	13.85	4.12	4.79	18.03	17.23	12.04
ISO viscosity grade	VG 22	VG 68	-	VG 22	VG 150	VG 150	VG 68
Viscosity index (VI)	188	199	259	160	121	132	184
Flash point (°C)	152	116	140	148	98	102	118
Pour point (°C)	11	−8	7	−6	−9	−4	6
Decomposition onset temperature (°C)	362	384	345	326	401	392	367
Spin-coated film thickness (μm)	4.85	8.6	3.84	4.21	11.4	10.27	7.42
Viscosity–pressure coefficient, α @25°C ($\times 10^{-8}$ Pa^{-1})	1.63	1.69	1.29	1.7	2.24	2.2	1.77

The flash point is affected by the unsaturated carbon–carbon bonds in the fatty acid chain of an ester. It is also subjected to the numbers and the location of the double bond in the chain. The double bond in the fatty acid chain act as the active side for oxidation [46]. In other words, the flash point of the TMP ester is expected to be higher with increasing SA. Referring to Table 3, the palm TMP ester (46% SA) had the highest flash point while TMP-canola (2.7% SA) had the lowest flash point, corresponding to 152 °C and 98 °C. Contrary to the pour point characterisation, defined as the lowest temperature fluidity of a lubricant, a lower degree of SA gives a better cold flow behaviour [47]. The obtained pour point for each TMP ester was found to follow the expected trend, where the pour point of canola was much lower than that of palm and coconut. It is noted that the flash and pour points of the TMP esters remained undesirable compared to commercial engine lubricants (≈220 °C for flash point and ≈−50 °C for pour point) and hydraulic oil (≈−20 °C for pour point). Thus, it can be surmised that the derived TMP esters would need further enhancements through flash point additives and pour point depressants.

The friction force measured for the derived vegetable-oil-derived TMP esters is given in Figure 4. The friction forces changed linearly with the applied loads for all the TMP esters. It is noted that the determination coefficient (R^2) for the friction forces as a function of load was above 0.85 when considering a linear regression. Such a trend followed the ones expected of boundary lubrication. However, the coefficient of friction (slope of the friction curve) varied when the sliding speed was increased. Another noticeable trend presented in Figure 4 is the vertical shift or offset of the measured friction force with the increasing sliding speeds. The offset indicated a varying intercept at the friction force axis. The observed nonzero intercept, also known as a dynamical friction parameter, could result from the TMP ester molecules adsorbing on the rubbing surfaces to form a boundary film [48], generating an interfacial shear resistance that needs to be overcome to sustain the sliding action. Overall, the frictional property of the palm TMP ester improved with higher speeds, leading to a lower friction coefficient than the other TMP esters. Such an improvement in boundary lubricity could result from the balanced saturation and monounsaturation levels of palm TMP ester, not seen in other TMP esters.

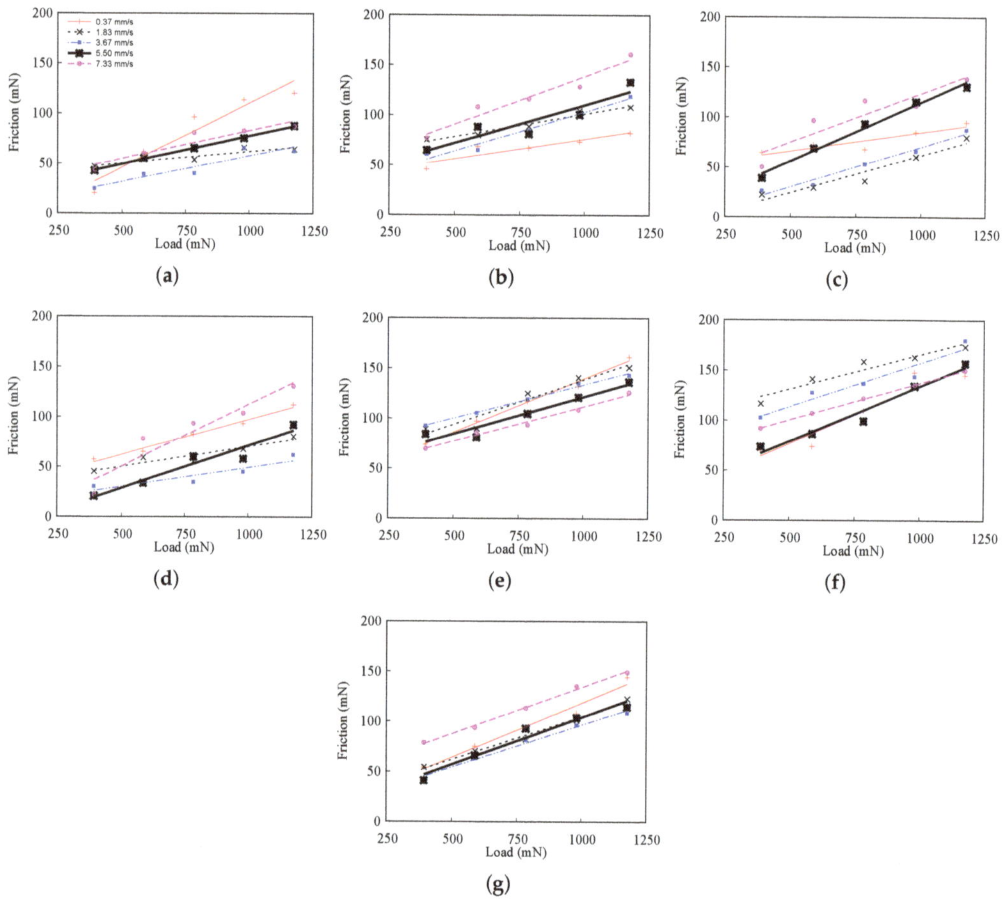

Figure 4. Friction force for vegetable-oil-derived TMP esters at varied loads and speeds. (**a**) Palm. (**b**) Olive. (**c**) Coconut. (**d**) Grapeseed–coconut. (**e**) Canola. (**f**) Canola–sunflower. (**g**) Canola–palm–soybean.

Figure 5 plots the coefficient of friction for the TMP esters against the lambda parameter (λ). It is noted that the coefficient of friction was taken as the slope of the measured friction force against the load provided in Figure 4 at its respective speed. An average λ was taken upon observing that the ratio was relatively constant for each load at its respective sliding speed. It is shown that all the TMP esters operated in mixed and boundary lubrication regimes ($\lambda < 1.15$). Interestingly, the coefficient of friction for all the TMP esters, except for the olive and coconut TMP esters, exhibited a Stribeck like property. The coefficient of friction dropped with increasing λ values until reaching a minimum before increasing slightly or saturating at higher λ values. Such a behaviour could result from an increased hydrodynamic effect even with a fixed amount of lubricant supply, encouraging the contact to transition from boundary to mixed lubrication regimes. The coefficient of friction for the olive and coconut TMP esters was observed to keep increasing with larger λ values. Such a trend indicated that these TMP esters might not sustain the film under a high shear stress, potentially attributed to the molecules' lesser ability to adsorb to the wear disks. It is noted that the coefficient of friction for all TMP esters was in the range of 0.03 and 0.14, much lower than the measured coefficient for a dry contact (0.34), indicating mixed and boundary lubrication regimes that were consistent with the λ values.

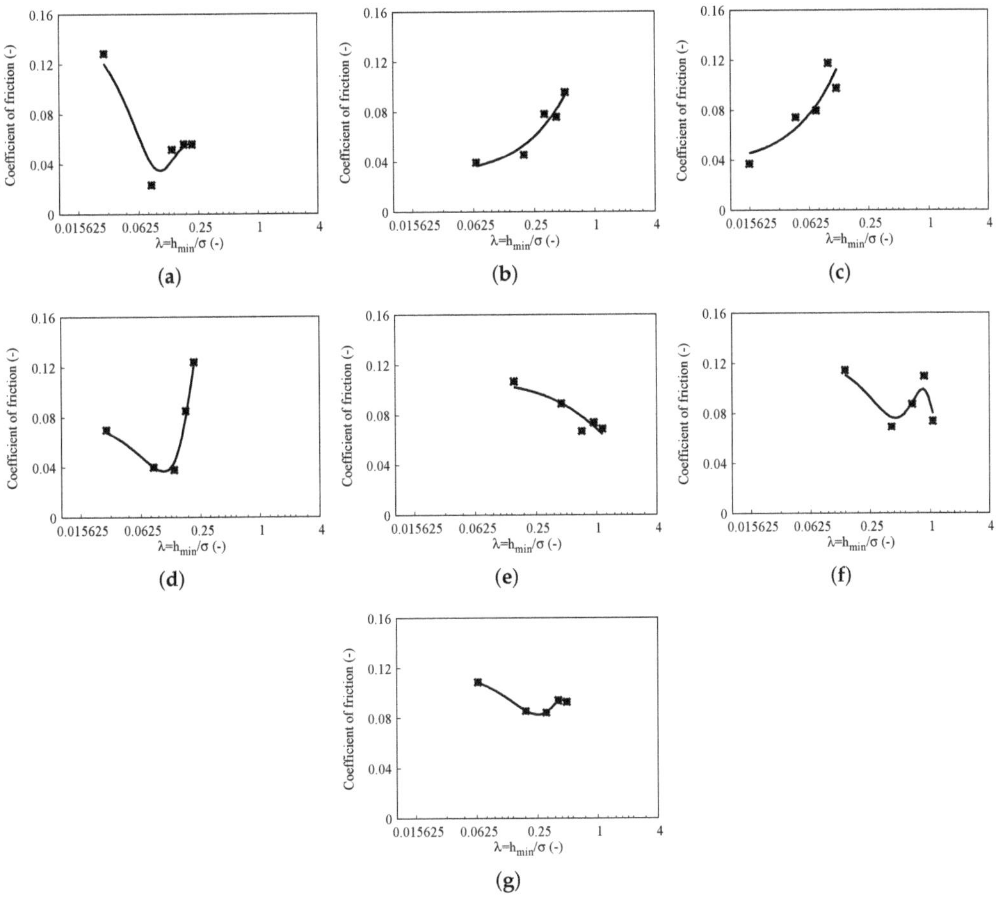

Figure 5. Coefficient of friction for vegetable-oil-derived TMP esters against the lambda parameter ($\lambda = h_{min}/\sigma$). (**a**) Palm. (**b**) Olive. (**c**) Coconut. (**d**) Grapeseed–coconut. (**e**) Canola. (**f**) Canola–sunflower. (**g**) Canola–palm–soybean. (Note: The dots represent the experiment data, while the lines represent the fitted trend line.)

GEP Model for TMP Ester Boundary Lubricity

The measured friction forces for the vegetable-oil-derived TMP esters were demonstrated to vary among each other. Considering the tested loads and speeds were fixed, such variation could be due to the influence of the fatty acid composition. However, such a correlation can only be qualitatively observed via the measured friction or coefficient of friction plots, but it remains challenging to quantify them mathematically. Even when it is possible to correlate them using typical regression models, the lack of generalisation capability limits the usage of the obtained correlations. Therefore, as a first approximation, the present study used the GEP model to describe the friction force as a function of the fatty acid composition of the TMP esters. Figure 6 illustrates the values for the training, validation and testing subsets, randomly assigned using a stratified sampling technique. The values for the validation and testing subsets were shown to sufficiently cover the range of data points represented by the training set. The GEP used the training subset to generate subexpression trees for the model, while the validation and testing subsets were adopted independently to validate the generated GEP model.

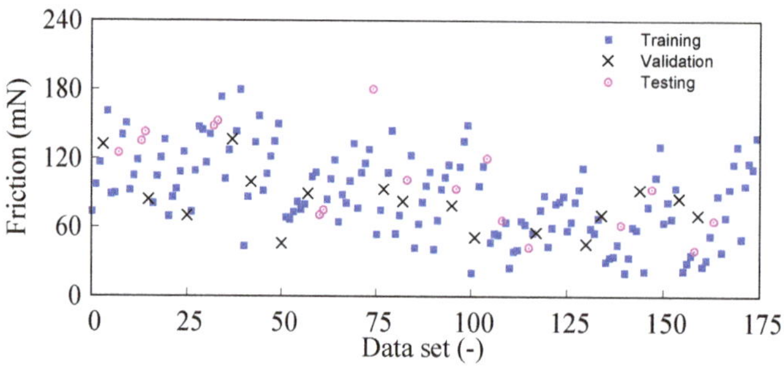

Figure 6. GEP model data from vegetable-oil-derived TMP ester friction tests for training, validation and testing sets.

The GEP model generated for the TMP ester friction is represented by six subexpression trees, each representing a gene from the GEP, in Figures 7 and 8. Each subexpression tree forms a term in the expression produced by the GEP model. Translating the subexpression trees would result in a correlation for the friction force that follows Equation (2). The friction force, F_f (mN), can be predicted using the following empirical equation:

$$F_f = A + B + C + D + log(E) + \frac{F_1}{F_2} \quad (3)$$

where

$A = \frac{1.0}{tan^{-1}(tan^{-1}(d_3 + d_0/d_1^2) - tanh(d_0)};$

$B = d_3 + 7.015 + \sqrt[3]{tan^{-1}(d_3 - d_0 + 29.447) \times d_2}$

$C = \sqrt[3]{d_2^2} - 6.21 + \sqrt[3]{(d_1 - d3 - 7.647) \times (d_2 + d_1)};$

$D = 5.072 + tanh(log(d_1 \times d_3)) \times \sqrt[3]{-(20.438 + d_0) \times d_1^2};$

$E = e^{d_3^2} + \frac{2.787 \times 10^9}{d_0};$

$F_1 = d_1 \times \left(\frac{1.743}{d_0 - d_1}\right);$

$F_2 = tan^{-1}\left(\frac{-4.338}{d_3} + \frac{d_1}{8.083}\right).$

The terms A to E are taken from subexpression trees one to five, while the terms F_1 and F_2 are from subexpression tree six6. The terms d_0, d_1, d_2 and d_3 refer to SA (%), MU (%), load (mN) and speed (m/s), respectively.

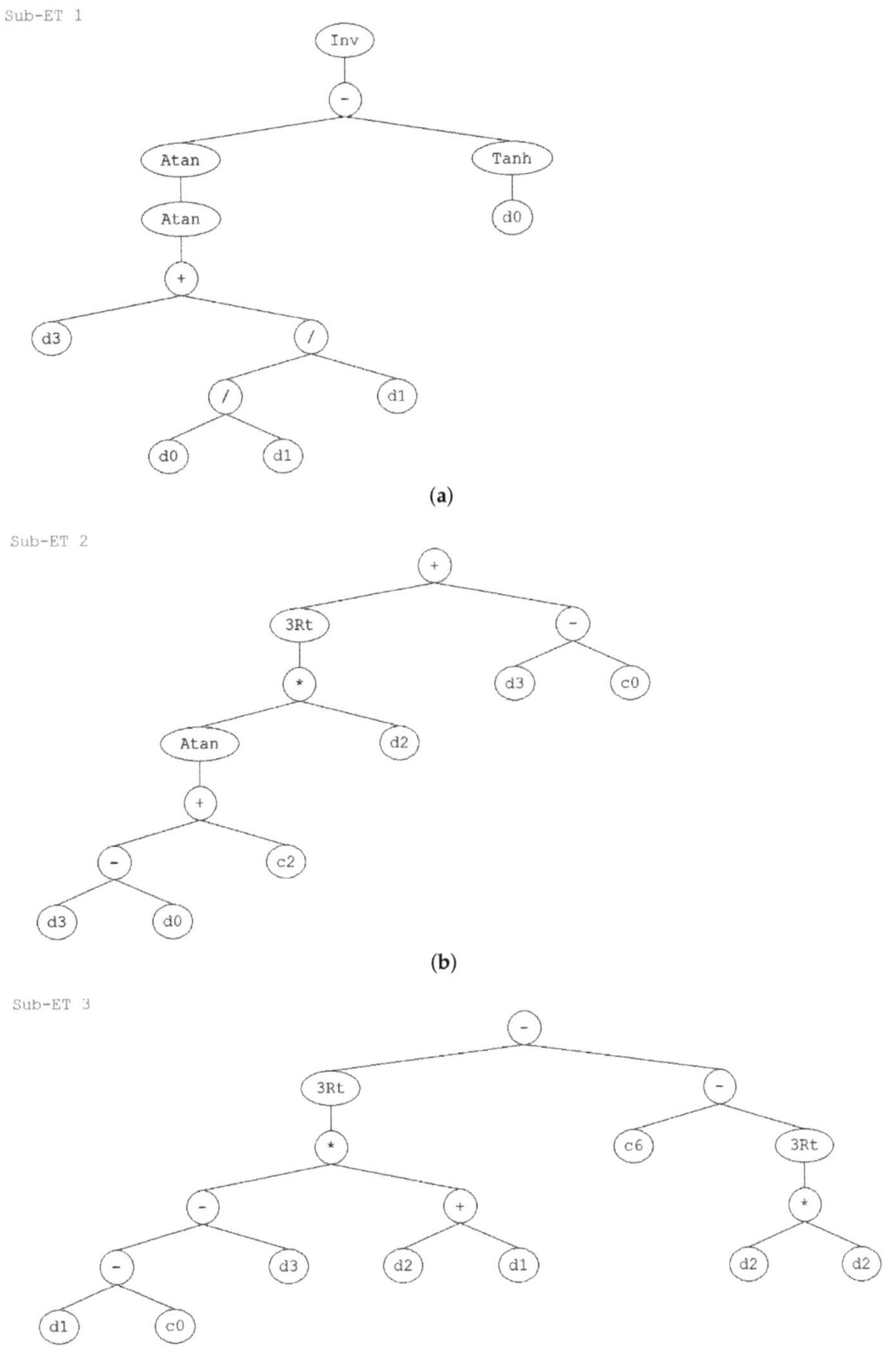

Figure 7. Subexpression trees 1 to 3 of the generated GEP model for vegetable-oil-derived TMP ester boundary lubricity. (**a**) Subexpression tree 1. (**b**) Subexpression tree 2. (**c**) Subexpression tree 3. (Note: "3Rt" refers to cubic root, "+" refers to addition, "-" refers to subtraction and "/" refers to division).

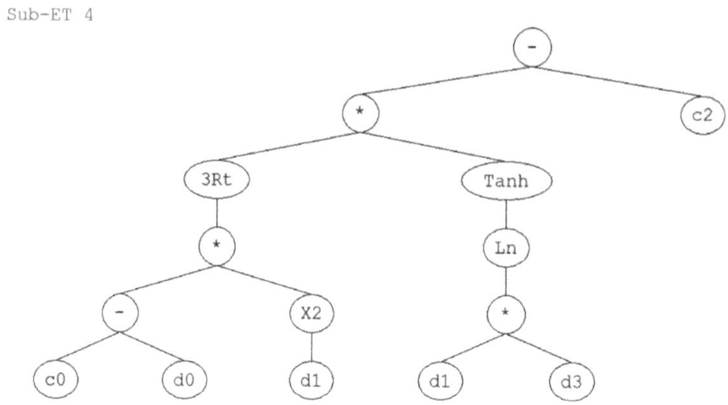

(a)

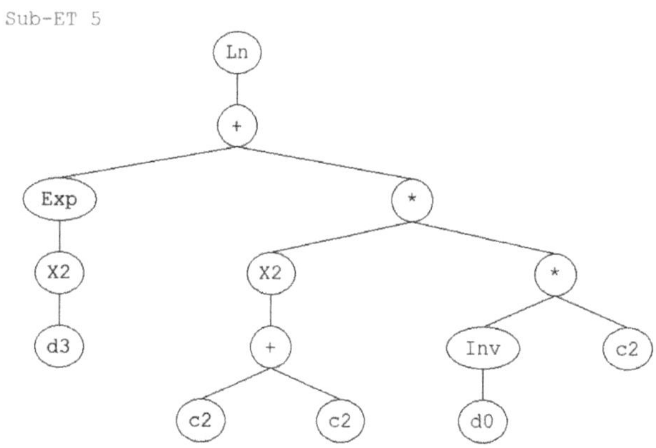

(b)

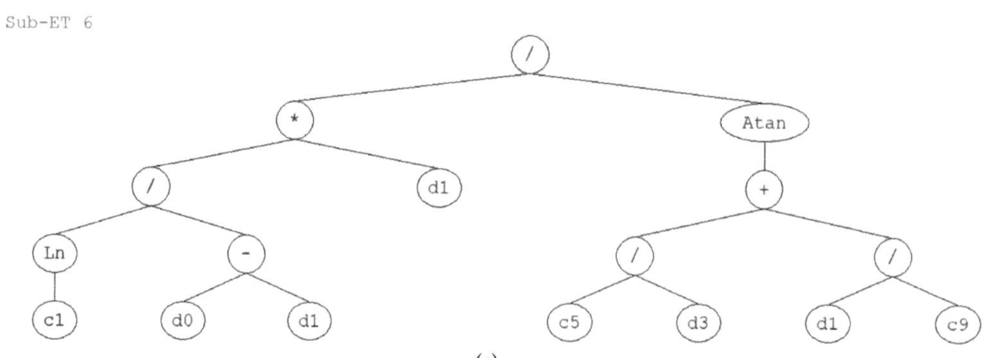

(c)

Figure 8. Subexpression trees 4 to 6 of the generated GEP model for vegetable-oil-derived TMP ester boundary lubricity. (**a**) Subexpression tree 4. (**b**) Subexpression tree 5. (**c**) Subexpression tree 6. (Note: 3Rt refers to cubic root, "+" refers to addition, "-" refers to subtraction and "/" refers to division).

Figure 9 gives the friction force comparison between the experiment (target) and model. The absolute error values are also provided in the exact figure. It can be observed that the model predicted friction force trends that followed the experimentally measured values for the training, validation and testing subsets. Table 4 summarises the statistical parameters for evaluating the GEP model performance. The determination coefficient (R^2) for the training, validation and testing subsets were 0.858, 0.824 and 0.916, respectively. On the other hand, the correlation coefficients (R) were 0.926, 0.908 and 0.957 for the training, validation and training subsets, respectively. These values were larger than 0.8 [34], indicating a strong correlation between experiment and model. The values for $RMSE$, MAE, RSE and $RRMSE$ were fairly similar among the training, validation and testing sets, indicating good generalisation capability of the model. Along with the near zero performance index, ρ, the generalisation capability of the generated GEP model was shown to be statistically reliable. The OBF values for the validation and testing subsets were 0.078 and 0.076, respectively. The near-zero values for OBF also indicated there were no overfitting issues [34].

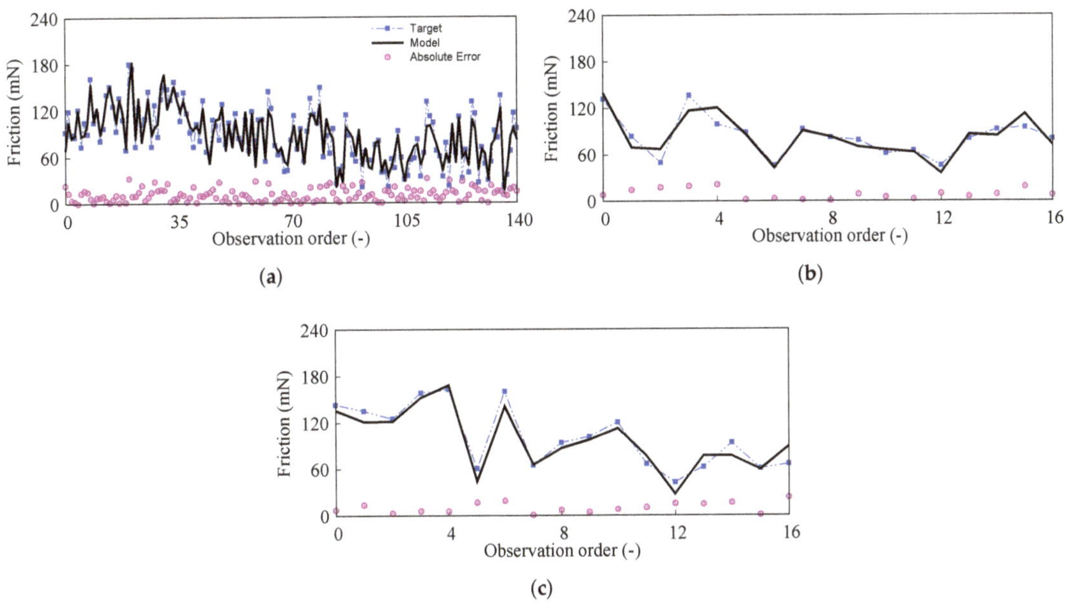

Figure 9. Friction force comparison between experiment (target) and GEP model for vegetable-oil-derived TMP esters. (**a**) Training subset. (**b**) Validation subset. (**c**) Testing subset.

Table 4. Statistical parameters of the GEP model.

Parameters	Training	Validation	Testing
$RMSE$	13.567	11.173	12.049
MAE	10.920	9.047	10.161
RSE	0.143	0.201	0.093
$RRMSE$	0.154	0.135	0.119
R	0.926	0.908	0.957
R^2	0.858	0.824	0.916
ρ	0.080	0.071	0.061
OBF	-	0.078	0.076

The GEP model was also externally validated using other statistical parameters as highlighted by Iqbal et al. [30]. The external validation's purpose was to evaluate the GEP model's generalisation capability further. Table 5 tabulated the statistical parameters adopted in the present study for the external validation of the GEP model. The slopes of the regression lines, k and k', for the training, validation and tests subsets were 0.976, 0.998 and 0.957, respectively. The values, close to unity, verified the correctness of the correlation produced by the GEP model [35]. The squared correlation coefficient through the origins ($R_o'^2$) and the coefficient between the experiment and model (R_o^2) were close to unity for all sets of data. Such trends strongly indicated that the generated GEP model had a statistically reliable generalisation capability and was not merely a correlation. More importantly, it can be surmised that the GEP model possessed a high generalisation capacity and excellent ability to predict reliable outcomes for unseen data or values.

Table 5. Statistical parameters of the GEP model for external validation.

Parameters	Criteria	Training	Validation	Testing
k	$0.85 < k < 1.15$	0.976	0.998	0.957
k'	$0.85 < k' < 1.15$	1.003	0.986	1.033
R_o^2	$R_o^2 \cong 1$	0.973	0.999	0.894
$R_o'^2$	$R_o'^2 \cong 1$	0.999	0.988	0.933

The variable importance level for the GEP model is also highlighted in Figure 10 along with the influence of each variable on the friction of the derived vegetable-oil-derived TMP esters. The GEP model highlighted that the load (38% contribution) was an essential variable, where increasing the load applied to the contact increased the friction force of the vegetable-oil-derived TMP esters. The saturation level, SA, was the second most important variable at 29%, where higher SA levels would result in a friction reduction. On the contrary, the GEP model indicated that the MU level (21%) and sliding speed (12%) were less significant in affecting the TMP ester friction force than load and SA. The lesser significance of the sliding speed from the GEP model was an expected trend as this corroborates the characteristic of boundary lubrication following the Stribeck curve, where the sliding speed has little influence on the boundary friction. Referring to Figure 10e, the friction force gradually increased with the sliding speed. The significance could be amplified further when the sliding speed increases beyond the range selected for the present study, potentially attributing to the growing hydrodynamic effect. However, it can be said that the present GEP model could be less accurate at higher sliding speeds, especially when the operating lubrication regime transitions more towards fluid film lubrication, attributed to the lack of information on lubrication regimes outside of the ones provided for the present model.

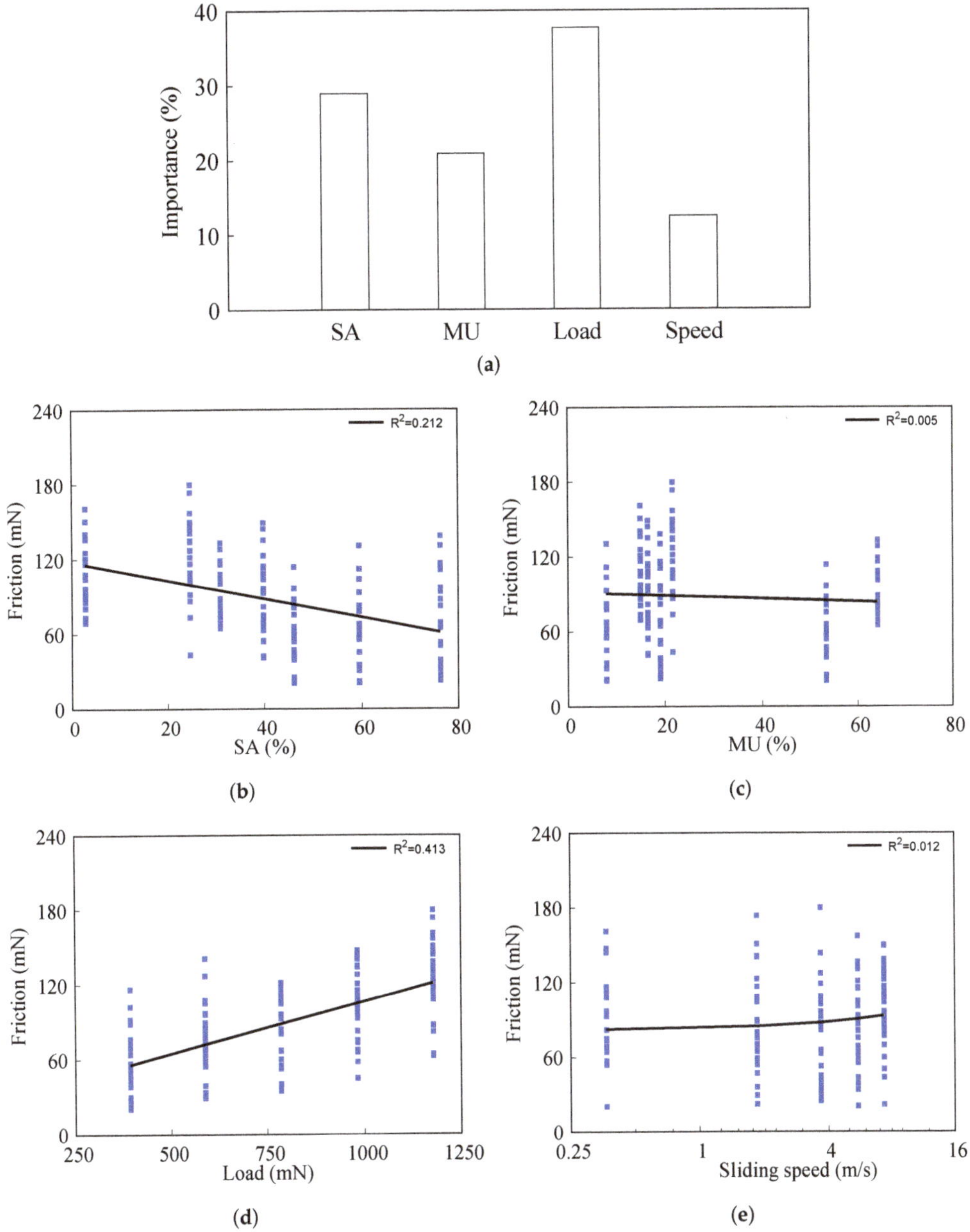

Figure 10. GEP model correlation between input variables and friction force for vegetable-oil-derived TMP esters. (**a**) Variable importance. (**b**) Saturation level (SA). (**c**) Monounsaturation level (MU). (**d**) Load. (**e**) Speed.

4. Conclusions

The present study synthesised TMP esters from different vegetable oil feedstocks with varying fatty acid profiles using a two-stage transesterification process. Except for the coconut TMP ester, the TMP esters could be mapped to ISO VG 22 (palm and grapeseed–

canola TMP esters), ISO VG 68 (olive and canola–palm–soybean TMP esters) and ISO VG 150 (canola and canola–sunflower TMP esters). The VI values of the TMP esters were comparable to typical engine lubricants. Lower VI values could be attributed to higher polyunsaturation levels while the opposite could be achieved with higher saturation levels.

The study then determined the boundary lubricity of vegetable-oil-derived TMP esters using a purpose-built ball-on-disc tribometer. The TMP esters were spin-coated on stainless steel wear discs, allowing the friction test to be carried out in a boundary lubrication regime. The coefficient of friction for all TMP esters, except olive and coconut TMP esters, exhibited a Stribeck like trend, potentially indicating the transition of operating lubrication regimes from boundary to mixed. Among the studied TMP esters, the palm TMP ester exhibited improved frictional performance with higher speeds, potentially due to its more balanced saturation and monounsaturation levels of fatty acid profile compared to the other tested TMP esters.

A gene expression programming (GEP) was adopted to model the boundary lubricity of the TMP esters. A set of simple and explicit equations was produced to describe the boundary lubricity of the TMP ester considering the fatty acid composition (saturation and monounsaturation levels), load and speed. The GEP model was empirical and agreed well with the measured friction force. The statistical evaluation of the GEP model, including an external validation, demonstrated that the model had a high generalisation and prediction capability. The model also showed that the friction force for the TMP esters decreased with higher saturation levels. On the contrary, the lesser influence of the speed followed the characteristics of boundary lubrication, where hydrodynamic action is negligible. Thus, the GEP model is expected to provide a fundamental and empirical platform for further studies to optimise the boundary tribological properties of vegetable-oil-based TMP esters, encouraging the widespread adoption of such a biolubricant as an alternative to conventional lubricants.

Author Contributions: Conceptualization, C.T.L., W.W.F.C., J.-H.N. and C.T.C.; data curation, C.T.L., M.B.L. and K.J.W.; formal analysis, C.T.L., W.W.F.C. and J.-H.N.; funding acquisition, W.W.F.C.; investigation, C.T.L. and M.B.L.; methodology, C.T.L. and J.-H.N.; project administration, W.W.F.C. and K.J.W.; resources, W.W.F.C. and K.J.W.; supervision, W.W.F.C., J.-H.N. and C.T.C.; validation, C.T.L. and M.B.L.; visualization, C.T.L. and W.W.F.C.; writing—original draft, C.T.L. and W.W.F.C.; writing—review and editing, W.W.F.C., J.-H.N. and C.T.C. All authors have read and agreed to the published version of the manuscript.

Funding: This research study was funded by the UTM Fundamental Research Grant, Universiti Teknologi Malaysia (UTM) under project no. Q.J130000.3851.22H02 (PY/2015/04925).

Data Availability Statement: Not applicable.

Acknowledgments: The authors would like to acknowledge the technical advice on machine learning techniques provided by Kuan Yew Wong from the Faculty of Mechanical Engineering, Universiti Teknologi Malaysia.

Conflicts of Interest: The authors declare no conflict of interest.

Appendix A. Input Parameters for GEP Algorithm

Parameters	Settings
General parameters	
Number of chromosomes	75, 150, 225
Head size	6, 8, 10
Number of genes	3, 6, 9
Linking function	Addition
Fitness function	Enhanced RSE
Function set	+, -, ×, ÷, exp, natural logarithm, power of 2, cube root, arctangent, hyperbolic tangent
Genetic operators	
Strategy	Optimal evolution
Mutation rate	0.00138
Inversion rate	0.00546
IS transportation rate	0.00546
RIS transportation rate	0.00546
One-point recombination rate	0.00277
Two-point recombination rate	0.00277
Gene recombination rate	0.00277
Gene transposition rate	0.00277
Numerical constants	
Constants per gene	10
Data type	Floating point
Upper-bound value	−10
Lower-bound value	10

Appendix B. Statistical Parameters

The statistical parameters used in the present study are summarised below:

$$RMSE = \sqrt{\left(\frac{\sum_{i=1}^{n}(e_i - m_i)^2}{n}\right)}$$

$$MAE = \frac{\sum_{i=1}^{n}|e_i - m_i|}{n}$$

$$RSE = \frac{\sum_{i=1}^{n}(m_i - e_i)^2}{\sum_{i=1}^{n}(\bar{e} - e_i)^2}$$

$$RRMSE = \frac{1}{|\bar{e}|} \times RMSE$$

$$R = \frac{\sum_{i=1}^{n}(e_i - \bar{e})(m_i - \bar{m})}{\sqrt{\sum_{i=1}^{n}(e_i - \bar{e})^2(m_i - \bar{m})^2}}$$

$$R^2 = R \times R$$

$$\rho = \frac{RRMSE}{1 + R}$$

$$k = \sqrt{\left(\frac{\sum_{i=1}^{n}(e_i - m_i)^2}{e_i^2}\right)}$$

$$k' = \sqrt{(\frac{\sum_{i=1}^{n}(e_i - m_i)^2}{m_i^2})}$$

$$R_o'^2 = 1 - \frac{\sum_{i=1}^{n}(m_i - e_i^o)^2}{\sum_{i=1}^{n}(m_i - m_i^o)^2}$$

$$R_o^2 = 1 - \frac{\sum_{i=1}^{n}(e_i - m_i^o)^2}{\sum_{i=1}^{n}(e_i - e_i^o)^2}$$

$$e_i^o = k \times m_i$$

$$m_i^o = k' \times e_i$$

$$OBF = \frac{n_T - n_V}{n} \times \rho_T + 2\frac{n_v}{n}\rho_v$$

The terms e and $\bar{e}$ refer to the experiment and its average values. In comparison, the terms m and $\bar{m}$ refer to the model and its average values, with the subscript i being the output value for the experiment and model. On the other hand, the term n is the number of values with subscripts T and V referring to training and validation subsets.

References

1. Research and Markets. Biolubricants—Global Market Trajectory & Analytics. 2022. Available online: https://www.researchandmarkets.com/reports/1824122/biolubricants_global_market_trajectory_and (accessed on 2 November 2022).
2. Singh, A.K.; Gupta, A.K. Metalworking fluids from vegetable oils. *J. Synth. Lub.* **2006**, *23*, 167–176. [CrossRef]
3. Fox, N.J.; Tyrer, B.; Stachowiak, G.W. Boundary lubrication performance of free fatty acids in sunflower oil. *Tribol. Lett.* **2004**, *16*, 275–281. [CrossRef]
4. Edla, S.; Thampi, A.D.; Prasannakumar, P.; Rani, S. Evaluation of physicochemical, tribological and oxidative stability properties of chemically modified rice bran and karanja oils as viable lubricant base stocks for industrial applications. *Tribol. Int.* **2022**, *2022*, 107631. [CrossRef]
5. Owuna, F.J.; Dabai, M.U.; Sokoto, M.A.; Dangoggo, S.M.; Bagudo, B.U.; Birnin-Yauri, U.A.; Hassan, L.G.; Sada, I.; Abubakar, A.L.; Jibrin, M.S. Chemical modification of vegetable oils for the production of biolubricants using trimethylolpropane: A review. *Egypt. J. Petrol.* **2020**, *29*, 75–82. [CrossRef]
6. Zulkifli, N.W.M.; Azman, S.S.N.; Kalam, M.A.; Masjuki, H.H.; Yunus, R.; Gulzar, M. Lubricity of bio-based lubricant derived from different chemically modified fatty acid methyl ester. *Tribol. Int.* **2016**, *93*, 555–562. [CrossRef]
7. Narayanasarma, S.; Kuzhiveli, B.T. Evaluation of lubricant properties of polyolester oil blended with sesame oil-An experimental investigation. *J. Clean. Prod.* **2021**, *281*, 125347. [CrossRef]
8. Aziz, N.A.M.; Yunus, R.; Hamid, H.A.; Ghassan, A.A.K.; Omar, R.; Rashid, U.; Abbas, Z. An acceleration of microwave-assisted transesterification of palm oil-based methyl ester into trimethylolpropane ester. *Sci. Rep.* **2020**, *10*, 1–17.
9. Attia, N.K.; El-Mekkawi, S.A.; Elardy, O.A.; Abdelkader, E.A. Chemical and rheological assessment of produced biolubricants from different vegetable oils. *Fuel* **2020**, *271*, 117578. [CrossRef]
10. Carvalho, W.C.A.; Luiz, J.H.H.; Fernandez-Lafuente, R.; Hirata, D.B.; Mendes, A.A. Eco-friendly production of trimethylolpropane triesters from refined and used soybean cooking oils using an immobilized low-cost lipase (Eversa>® Transform 2.0) as heterogeneous catalyst. *Biomass Bioener* **2021**, *155*, 106302. [CrossRef]
11. Gul, M.; Zulkifli, N.W.M.; Masjuki, H.H.; Kalam, M.A.; Mujtaba, M.A.; Harith, M.H.; Syahir, A.Z.; Ahmed, W.; Farooq, A.B. Effect of TMP-based-cottonseed oil-biolubricant blends on tribological behavior of cylinder liner-piston ring combinations. *Fuel* **2020**, *278*, 118242. [CrossRef]
12. Guimarey, M.J.G.; Goncalves, D.E.P.; del Río, J.M.L.; Comuñas, M.J.P.; Fernández, J.; Seabra, J.H.O. Lubricant properties of trimethylolpropane trioleate biodegradable oil: High pressure density and viscosity, film thickness, Stribeck curves and influence of nanoadditives. *J. Mol. Liq.* **2021**, *335*, 116410. [CrossRef]
13. Elmelawy, M.S.; El-Meligy, A.; Mawgoud, H.A.; Morshedy, A.S.; Hanafy, S.A.; El-sayed, I.E. Synthesis and kinetics study of trimethylolpropane fatty acid triester from oleic acid methyl ester as potential biolubricant. *Biomass Convers. Biorefinery* **2021**. [CrossRef]

14. Lee, C.T.; Lee, M.B.; Mong, G.R.; Chong, W.W.F. A bibliometric analysis on the tribological and physicochemical properties of vegetable oil–based bio-lubricants (2010–2021). *Environ. Sci. Pollut. Environ.* **2022**, *29*, 56215–56248. [CrossRef]
15. Rahim, E.A.; Amiril, S.A.S.; Mohid, Z.; Syahrullail, S. Tribological evaluation on vegetable oils based trimethylolpropane (TMP) ester as sustainable metalworking fluids in machining applications. In Proceedings of the 7th International Conference Mechanical, Manufacturing Engineering, Yogyakarta, Indonesia 2016, August 1-3; pp. 1–8.
16. Verma, P.; Sharma, M.P. Review of process parameters for biodiesel production from different feedstocks. *Renew. Sustain. Energy Rev.* **2016**, *62*, 1063–1071. [CrossRef]
17. Zainal, N.A.; Zulkifli, N.W.M.; Gulzar, M.; Masjuki, H.H. A review on the chemistry, production, and technological potential of bio-based lubricants. *Renew. Sustain. Energy Rev.* **2018**, *82*, 80–102. [CrossRef]
18. Biresaw, G.; Bantchev, G.B. Pressure viscosity coefficient of vegetable oils. *Tribol. Lett.* **2013**, *49*, 501–512. [CrossRef]
19. Fasina, O.O.; Hallman, H.; Craig-Schmidt, M.; Clements, C. Predicting temperature-dependence viscosity of vegetable oils from fatty acid composition. *J. Am. Oil Chem. Soc.* **2006**, *83*, 899–903. [CrossRef]
20. Kim, J.; Kim, D.N.; Lee, S.H.; Yoo, S.; Lee, S. Correlation of fatty acid composition of vegetable oils with rheological behaviour and oil uptake. *Food Chem.* **2010**, *118*, 398–402. [CrossRef]
21. Verma, P.; Sharma, M.P.; Dwivedi, G. Evaluation and enhancement of cold flow properties of palm oil and its biodiesel. *Energy Rep.* **2016**, *2*, 8–13. [CrossRef]
22. Siniawski, M.T.; Saniei, N.; Adhikari, B.; Doezema, L.A. Influence of fatty acid composition on the tribological performance of two vegetable-based lubricants. *J. Synth. Lub.* **2007**, *24*, 101–110. [CrossRef]
23. Rajasozhaperumal, G.; Kannan, C. Influence of Fatty Acid Composition on the Tribological Performance of Methyl Esters Under Boundary Lubrication Regime. *Arab. J. Sci. Eng.* **2022**. [CrossRef]
24. Hamdan, S.H.; Chong, W.W.F.; Ng, J.H.; Ghazali, M.J.G.; Wood, R.J.K. Influence of fatty acid methyl ester composition on tribological properties of vegetable oils and duck fat derived biodiesel. *Tribol. Int.* **2017**, *113*, 76–82. [CrossRef]
25. Rosenkranz, A.; Marian, M.; Profito, F.J.; Aragon, N.; Shah, R. The use of artificial intelligence in tribology—A perspective. *Lubricants* **2020**, *9*, 2. [CrossRef]
26. Marian, M.; Tremmel, S. Current trends and applications of machine learning in tribology? A review. *Lubricants* **2021**, *9*, 86. [CrossRef]
27. Durak, E.; Salman, Ö.; Kurbanoğlu, C. Analysis of effects of oil additive into friction coefficient variations on journal bearing using artificial neural network. *Ind. Lub. Tribol.* **2008**, *60*, 309–316. [CrossRef]
28. Bhaumik, S.; Mathew, B.R.; Datta, S. Computational intelligence-based design of lubricant with vegetable oil blend and various nano friction modifiers. *Fuel* **2019**, *241*, 733–743. [CrossRef]
29. Sattari Baboukani, B.; Ye, Z.; G Reyes, K.; Nalam, P.C. Prediction of nanoscale friction for two-dimensional materials using a machine learning approach. *Tribol. Lett.* **2020**, *68*, 1–14. [CrossRef]
30. Iqbal, M.F.; Liu, Q.; Azim, I.; Zhu, X.; Yang, J.; Javed, M.F.; Rauf, M. Prediction of mechanical properties of green concrete incorporating waste foundry sand based on gene expression programming. *J. Hazard. Mater.* **2020**, *384*, 121322. [CrossRef]
31. Sebaaly, H.; Varma, S.; Maina, J.W. Optimizing asphalt mix design process using artificial neural network and genetic algorithm. *Construct. Build. Mater.* **2018**, *168*, 660–670. [CrossRef]
32. Ferreira, C. Gene expression programming: A new adaptive algorithm for solving problems. *arXiv* **2001**, arXiv:cs/0102027.
33. Faradonbeh, R.S.; Armaghani, D.J.; Monjezi, M.; Mohamad, E.T. Genetic programming and gene expression programming for flyrock assessment due to mine blasting. *Int. J. Rock. Mech. Mining Sci.* **2016**, *88*, 254–264. [CrossRef]
34. Gandomi, A.H.; Roke, D.A. Assessment of artificial neural network and genetic programming as predictive tools. *Adv. Eng. Softw.* **2015**, *88*, 63–72. [CrossRef]
35. Aslam, F.; Farooq, F.; Amin, M.N.; Khan, K.; Waheed, A.; Akbar, A.; Javed, M.F.; Alyousef, R.; Alabdulijabbar, H. Applications of gene expression programming for estimating compressive strength of high-strength concrete. *Adv. Civ. Eng.* **2020**, *2020*. [CrossRef]
36. Ren, L.; Wang, N.; Pang, W.; Li, Y.; Zhang, G. Modeling and monitoring the material removal rate of abrasive belt grinding based on vision measurement and the gene expression programming (GEP) algorithm. *Int. J. Adv. Manufact. Tech.* **2022**, *120*, 385–401. [CrossRef]
37. Mahdinia, S.; Eskandari-Naddaf, H.; Shadnia, R. Effect of cement strength class on the prediction of compressive strength of cement mortar using GEP method. *Construct. Build. Mater.* **2019**, *198*, 27–41. [CrossRef]
38. Tariq, M.; Ali, S.; Ahmad, F.; Ahmad, M.; Zafar, M.; Khalid, N.; Khan, M.A. Identification, FT-IR, NMR (1H and 13C) and GC/MS studies of fatty acid methyl esters in biodiesel from rocket seed oil. *Fuel Process. Technol.* **2011**, *92*, 336–341. [CrossRef]
39. Yunus, R.; Lye, O.T.; Fakhru'l-Razi, A.; Basri, S. A simple capillary column GC method for analysis of palm oil-based polyol esters. *J. Am. Oil Chem. Soc.* **2002**, *79*, 1075–1080. [CrossRef]
40. Meng, F.; Han, H.; Ma, Z.; Tang, B. Effects of Aviation Lubrication on Tribological Performances of Graphene/MoS2 Composite Coating. *J. Tribol.* **2021**, *143*, 031401. [CrossRef]
41. Hamrock, B.J.; Dowson, D. Isothermal elastohydrodynamic lubrication of point contacts: Part IV–starvation results. *J. Lubrication Tech.* **1977**, *99(1)*, 15–23. 1977. [CrossRef]
42. Dunlop, P.; Smith, S. Estimating key characteristics of the concrete delivery and placement process using linear regression analysis. *Civ. Eng. Environ. Syst.* **2003**, *20*, 273–290. [CrossRef]

43. Mousavi, S.M.; Aminian, P.; Gandomi, A.H.; Alavi, A.H.; Bolandi, H. A new predictive model for compressive strength of HPC using gene expression programming. *Adv. Eng. Softw.* **2012**, *45*, 105–114. [CrossRef]
44. Gandomi, A.H.; Babanajad, S.K.; Alavi, A.H.; Farnam, Y. Novel approach to strength modeling of concrete under triaxial compression. *J. Mater. Civ. Eng.* **2012**, *24*, 1132–1143. [CrossRef]
45. Lee, C.T.; Lee, M.B.; Hamdan, S.H.; Chong, W.W.F.; Chong, C.T.; Zhang, H.; Chen, A.W.L. Trimethylolpropane trioleate as eco-friendly lubricant additive. *Eng. Sci. Tech. Int. J.* **2022**, *35*, 101068. [CrossRef]
46. Yuan, W.; Hansen, A.C.; Zhang, Q. Vapor pressure and normal boiling point predictions for pure methyl esters and biodiesel fuels. *Fuel* **2005**, *84*, 943–950. [CrossRef]
47. Folayan, A.J.; Anawe, P.A.L.; Aladejare, A.E.; Ayeni, A.O. Experimental investigation of the effect of fatty acids configuration, chain length, branching and degree of unsaturation on biodiesel fuel properties obtained from lauric oils, high-oleic and high-linoleic vegetable oil biomass. *Energy Rep.* **2019**, *5*, 793–806. [CrossRef]
48. He, Y.; Fujikawa, Y.; Zhang, H.; Fukuzawa, K.; Mitsuya, Y. Evaluations of tribological characteristics of PFPE lubricants on DLC surfaces of magnetic disks. *Tribol. Lett.* **2007**, *27*, 1–11. [CrossRef]

MDPI AG
Grosspeteranlage 5
4052 Basel
Switzerland
Tel.: +41 61 683 77 34

Lubricants Editorial Office
E-mail: lubricants@mdpi.com
www.mdpi.com/journal/lubricants

www.ingramcontent.com/pod-product-compliance
Lightning Source LLC
Chambersburg PA
CBHW062037220926
51850CB00010B/2576
* 9 7 8 3 7 2 5 8 5 2 0 5 5 *